SpringerBriefs in Electrical and Computer Engineering

Series editors
Woon-Seng Gan, School of Electrical and Electronic Engineering, Nanyang
Technological University, Singapore, Singapore
C.-C. Jay Kuo, University of Southern California, Los Angeles, CA, USA
Thomas Fang Zheng, Research Institute of Information Technology, Tsinghua
University, Beijing, China
Mauro Barni, Department of Information Engineering and Mathematics,
University of Siena, Siena, Italy

T0171957

SpringerBriefs present concise summaries of cutting-edge research and practical applications across a wide spectrum of fields. Featuring compact volumes of 50 to 125 pages, the series covers a range of content from professional to academic. Typical topics might include: timely report of state-of-the art analytical techniques, a bridge between new research results, as published in journal articles, and a contextual literature review, a snapshot of a hot or emerging topic, an in-depth case study or clinical example and a presentation of core concepts that students must understand in order to make independent contributions.

More information about this series at http://www.springer.com/series/10059

Iraj Sadegh Amiri • Mahdi Ariannejad

Introducing CTS (Copper-Tin-Sulphide) as a Solar Cell by Using Solar Cell Capacitance Simulator (SCAPS)

 Springer

Iraj Sadegh Amiri
Computational Optics Research Group,
Advanced Institute of Materials Science
Ton Duc Thang University
Ho Chi Minh City, Vietnam

Faculty of Applied Sciences
Ton Duc Thang University
Ho Chi Minh City, Vietnam

Mahdi Ariannejad
University of Malaya
Kuala Lumpur, Malaysia

ISSN 2191-8112 ISSN 2191-8120 (electronic)
SpringerBriefs in Electrical and Computer Engineering
ISBN 978-3-030-17394-4 ISBN 978-3-030-17395-1 (eBook)
https://doi.org/10.1007/978-3-030-17395-1

This Springer imprint is published by the registered company Springer Nature Switzerland AG
The registered company address is: Gewerbestrasse 11, 6330 Cham, Switzerland

Preface

Solar energy is gaining increasing popularity in this modern world. Solar energy is one of the preferred alternative sources of energy to substitute the fossil fuel energy as it is environmentally friendly. In this book, a thorough study on the performance of the CTS solar cell using Solar Cell Capacitance Simulator (SCAPS) is presented. The performance of the CTS solar cell was evaluated in terms of V_{oc}, J_{sc}, fill factor and efficiency. Structural parameter variation of CTS solar cell has been studied in terms of buffer layer and absorber layer thickness, bandgap and effect of temperature on total efficiency of the cell. Simulation results show that the efficiency of the solar cell decreases with the increased thickness of the CdS buffer layer. The highest efficiency of 20.36% is measured, when the buffer layer thickness is 10 nm. In terms of the CTS absorber layer thickness, it has been observed that the efficiency of the solar cell is increased by the increased thickness of absorber layer. The highest efficiency of 20.36% is measured, when the CTS thickness is 4 μm. Also it has been found that while the bandgap increases the efficiency of the solar cell decreases. In this research 0.9 eV bandgap resulting 11.58% cell efficiency, while by increasing the bandgap value the efficiency of cell increase and 1.25 bandgap resulting for 21.96% cell efficiency. In terms of temperature, efficiency of 20.36% results in 300 K temperature, and as the temperature increases cell efficiency decreases.

Ho Chi Minh City, Vietnam
Kuala Lumpur, Malaysia

Iraj Sadegh Amiri
Mahdi Ariannejad

Contents

List of Figures

List of Tables

List of Symbols

μ_n	Electron mobility
μ_p	Hole mobility
A	Diode's ideality factor
D_n	Electron diffusion coefficient
D_p	Hole diffusion coefficient
$E(\text{photon})$	Photon's energy in eV
E_c	Conduction band edge
E_f	Fermi level
E_g	Semiconductor energy bandgap
Eg_{back}	Back surface bandgap energy
Eg_{front}	Front surface bandgap energy
Eg_{min}	Minimum bandgap energy
E_T	Trap energy level
E_v	Valence band edge
$\mathcal{E}$	Electric field
$F(E)$	Fermi-Dirac distribution function
FF	Fill factor
G_{op}	Optical generation
h	Planck's constant
I	Current
I_0	Diode saturation current
I_D	Shockley diode current
I_L	Photogenerated current
I_{max}	Maximum current
I_{sc}	Short circuit current
I_{SH}	Shunt current
$J_{n,\,\text{diffusion}}$	Electron diffusion current
$J_{n,\,\text{drift}}$	Electron drift current
$J_{p,\,\text{diffusion}}$	Hole diffusion current
$J_{p,\text{drift}}$	Hole drift current

J_{total}	Total current
k	Boltzmann constant
K	Kelvin
L_{back}	Back grading width
L_{front}	Front grading width
m_e^*	Effective mass of electron
m_h^*	Effective mass of hole
n	Electron concentration in the conduction band
$N(E)$	Density of state function
N_A	Acceptor impurity concentration
N_c	Effective density of states in conduction band
N_D	Donor impurity concentration
n_i	Intrinsic carrier concentration at thermal equilibrium
n_o	Electron concentration in the conduction band at thermal equilibrium
n_t	Electron concentration at the trap energy level
N_v	Effective density of states in valence band
p	Hole concentration in the valence band
P_{in}	Input power
P_{max}	Maximum power point
p_o	Hole concentration in the valence band at thermal equilibrium
p_t	Hole concentration at the trap energy level
q	Electron's elementary charge
$R(x)$	Recombination rate function
R_s	Series resistance
R_{sh}	Shunt resistance
R^{SRH}	Rate of Shockley-Read-Hall recombination
T	Absolute temperature
V	Voltage
V_{bi}	Built-in potential
V_{max}	Maximum voltage
V_{oc}	Open circuit voltage
α	Absorption coefficient
ε_s	Dielectric constant
η	Conversion efficiency
λ	Wavelength
τ_n	Electron lifetime
τ_p	Hole lifetime
υ	Frequency
Ψ	Electrostatic potential

List of Nomenclature

Al	Aluminium
Al_2O_3	Aluminium oxide
Ar	Argon
Cd	Cadmium
$Cd(NH_3)_4^{2+}$	Tetramminecadmium
CdI_2	Cadmium iodide
CdO	Cadmium oxide
CdS	Cadmium sulphide
$CdSO_4$	Cadmium sulphate
CdTe	Cadmium telluride
CIGS	Copper indium gallium diselenide
CTS	Copper-tin-sulphide
Cu	Copper
$CuInSe_2$	Copper indium diselenide
$CuSe_2$	Copper diselenide
CZTS	Copper-zinc-tin-sulphide
F	Fluorine
Ga	Gallium
H_2	Hydrogen
H_2O	Water
In	Indium
In_2O_3	Indium oxide
In_2Se_3	Indium selenide
InS	Indium sulphide
ITO	Indium tin oxide
i-ZnO	Intrinsic zinc oxide
MgF_2	Magnesium fluoride
Mo	Molybdenum
O_2	Oxygen
O_3	Ozone
OH^-	Hydroxide

$SC(NH_2)_2$	Thiourea
Se	Selenium
Si	Silicon
SnO_2	Tin oxide
SnS	Tin monosulfide
SO_4^{2-}	Sulphate
ZnMgO	Zinc magnesium oxide
ZnO	Zinc oxide
ZnO:Al	Al_2O_3 doped zinc oxide
ZnS	Zinc sulphide

Chapter 1
Development of Solar Cell Photovoltaics: Introduction and Working Principles

1.1 Motivation

If we observe our environment to see what is the most abundant free resource available to us, the answer is solar energy. Hence, it is a natural alternative to fossil fuels such as oil, gas, and coal. Energy from fossil fuels and water is used to generate electricity on a large scale. Scarcity and wastage of these resources have made them expensive to use. Solar energy is a low-cost and safe alternative to those other fuels. There are many reasons to use solar energy, such as the expense of fossil fuels and biofuels, *global warming*, and developments in solar energy utilization.

1.2 A Brief History of Solar Cells

A solar cell, also known as a photovoltaic (PV) cell, harvests sunlight and transfers the energy into electricity by the photovoltaic effect. The term "photovoltaic" is based on the Greek word *phos* (meaning "light") and the word "voltaic" (meaning "electric"), which comes from the name of the Italian physicist Alessandro Volta, after whom the unit of electric potential, the volt, is named. The photovoltaic effect was discovered in 1839 by the French physicist Alexandre-Edmond Becquerel. More than 40 years later, in 1883, the first solar cell was built by an American, Charles Fritts, who coated selenium with a very thin layer of gold to form junctions, resulting in efficiency of only 1%. Then, in 1941—more than 100 years after the photovoltaic effect was discovered—Russell Ohl built the first silicon-doped solar cell. The modern age of solar power technology arrived in 1954 when Bell Laboratories, experimenting with semiconductors, accidentally found that silicon doped with certain impurities was very sensitive to light. Daryl Chapin, Calvin

© The Author(s), under exclusive license to Springer Nature Switzerland AG 2019
I. S. Amiri, M. Ariannejad, *Introducing CTS (Copper-Tin-Sulphide) as a Solar Cell by Using Solar Cell Capacitance Simulator (SCAPS)*, SpringerBriefs in Electrical and Computer Engineering, https://doi.org/10.1007/978-3-030-17395-1_1

Souther Fuller, and Gerald Pearson of Bell Laboratories then developed the first silicon solar cell capable of converting sunlight energy into enough power to run everyday electric equipment, with efficiency of about 5% [1]. Some of the first practical solar cells to be produced, with efficiency of around 6%, were used for satellites [2].

Nowadays, the solar cells on the market are mostly silicon wafer–based cells, which approach a theoretical efficiency value of 33% [1]. They are often considered "first-generation" solar cells. The most widely commercialized "second-generation" solar cells are cadmium telluride (CdTe)–based cells and copper indium gallium selenide (CIS or CIGS)–based cells. These materials are applied in a thin film on a conducting substrate such as coated glass or stainless steel. The success of second-generation solar cell benefits greatly from new manufacturing techniques such as chemical vapor deposition (CVD) and atomic layer deposition (ALD). These alternative manufacturing techniques were developed so that high-temperature processing could be reduced. It is believed that as the manufacturing process matures, the cost of production will depend more on the cost of the constituent materials. An alternative, kesterite Cu, has been found to be a good substitute for CIGS. "Third-generation" solar cells will improve on the electric performance of the second generation while maintaining low production costs. Current research on new solar cells is focused on dye-sensitized solar cells (DSCs), nanoparticle processing (for example, using quantum dots), etc.

1.3   Background

For millennia, man has constantly been on the hunt for ways in which he could produce energy. The evolution and advancements in energy production go back to the days when nomadic man invented fire by rubbing two stones together. In the modern world, almost every function involves various sources of energy, such as fossil fuels, natural gas, hydroelectricity, nuclear power, geothermal power, or solar power. Cost was not a criterion in the primitive world, when resources were in abundance. As man evolved, so did his needs for energy consumption.

With a present world population of more than 6.5 billion people and an estimate of 10 billion by 2055 [3] (Fig. 1.1), there is a serious task at hand to look for energy sources that, once exploited, will not alter the ecological balance. These sources are known as renewable sources of energy [4].

Overdependence on nonrenewable sources by man to date has pushed them to the limit of scarcity and, within a couple of decades, will push them to the brink of extinction. We have also witnessed their disadvantages such as pollution, irreversible damage to the environment, escalation of the price of fuel day by day, and, most importantly, the fact that they eventually run out. These situations can be avoided only if we are able to make the transition to renewable sources and can do so as soon as possible.

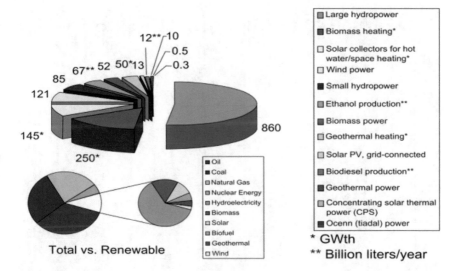

Fig. 1.1 Comparison of the various renewable energy production sources with each other and with total energy production. *GWth* gigawatts (thermal), *PV* photovoltaic

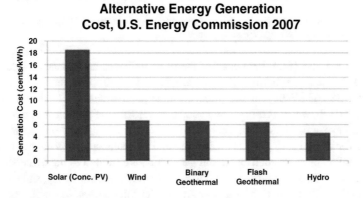

Fig. 1.2 Present price comparisons of different renewable energy sources, expressed in cents/kW/h. *Conc.* concentrated, *PV* photovoltaic

The two major limitations we face in synthesizing energy from renewable sources are cost and efficiency. The cost to generate electricity from coal is around US$0.04/kW/h and that from gas and oil is a little bit more: around US$0.08/kW/h. Figure 1.2 shows the current prices of electricity production by various *renewable energy sources*. Wind power, geothermal power, and hydroelectricity are priced within the range of US$0.05–0.07/kW/h, but solar (photovoltaic) power still remains expensive at around US$0.19/kW/h. With the extensive research going into each of these energy sources, the prices are estimated to drop in the next couple of years, as

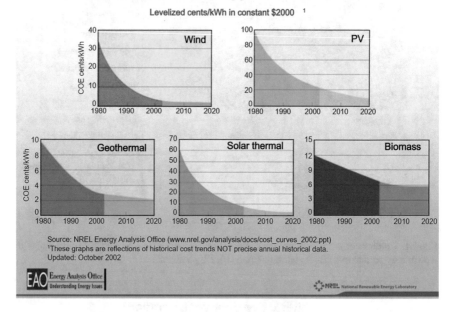

Fig. 1.3 Estimated drops in production costs (per kW/h) for different renewable sources of energy. *CCE* cost of conserved energy, *PV* photovoltaic

shown in Fig. 1.3 [5], where each of the plots indicates an exponential decrease. Wind, solar, and geothermal energy are expected to drop to very low prices of around US$0.02/kW/h, but the price of solar photovoltaics will still be around US$0.10/kW/h, which is quite high in comparison with the present coal pricing [6].

Solar energy is a renewable energy source, which is relatively inexpensive. The reasons for its popularity are its availability and its abundance; this natural resource is free, unlike biofuels such as oil and gas, which are limited to the ownership of a few countries. Those countries have a monopoly over these resources, and the prices of biofuels such as coal and oil fluctuate in the international market. Moreover, development of a biofuel *power plant* is an expensive project. Hence, if we work on developing *solar energy power plants*, we can save a sufficient amount of tax imposed on importing oil and other biofuels. All one needs is to deploy solar grids. However, a *solar power project* does require a minimum initial investment. The deployment of solar grids is scalable because the more solar grids we deploy, the more electricity we produce.

The use of fuels such as oil and gas in homes, cars, and industry has brought us the problem of global warming. The most extreme production of harmful gases such as carbon monoxide has destroyed the ozone layer; hence, we receive both harmless and harmful sunrays. The extreme pollution of our planet has disturbed the smooth functioning of our ecosystems. This has resulted in lower rainfall and dry weather.

Use of the sun to support industrial processes can help us overcome the worst situation of global warming. It can also help us to stop destroying our fertile land with harmful waste resulting from industrial processes. Governments all over the world are now supporting solar power projects, especially in developing countries. South Asian countries such as Pakistan, India, and Bangladesh are planning major solar power projects to utilize the benefits of solar energy.

1.4   Development and Research in the Field of Alternative Fuels

Research in the field of alternative fuels has helped scientists to create renewable energy projects. One such solar project, which was recently initiated in the Red Sea, is an experiment with solar ponds. The Red Sea is deep and rich in salt content. The presence of the salt is utilized to store solar heat in the sea. The temperature in the lower salt layers reaches 90 °C, which is sufficient for processing of heat and water.

Such developments and more in this field are economical; this is another reason why we should switch to alternative energy sources. Solar energy is helpful in generating tax-free electricity at home, which is why many end users have now deployed batteries that support solar electricity generation. Solar energy has many uses such as water treatment, cooking, and ventilation. Awareness about use of solar cookers in rural areas has also helped to decrease felling of trees for cooking fuel. Solar vehicles do not emit harmful gases as other vehicles do. Hence, all in all, the profusion and simplicity of solar energy has given us innumerable reasons to use it as a renewable energy source either alone or in combination with wind and geothermal energy sources.

1.5   Principles of Solar Operation

In semiconductors, the conduction bands are empty and are like insulators. However, because of the relatively small bandgap value ($E_g <3.5$ eV), electrons can be elevated to a conduction band through light or heat.

The operation of a photovoltaic cell requires three basic attributes:

1. Absorption of light, generating either electron–hole pairs or exceptions
2. Separation of charge carriers of opposite types
3. Separate extraction of those carriers to an external circuit

Incident photons with lower energy than E_g cannot excite electrons, while photons with sufficient energy to bridge the bandgap can lift electrons to the conduction band, with any excess energy being transferred to heat. This process is shown in Fig. 1.4 [7].

Fig. 1.4 Electrons are elevated from the valence band to the conduction band after absorbing the energy of incoming photons in a semiconductor

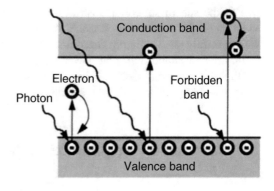

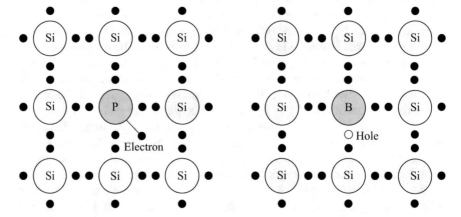

Fig. 1.5 n-Type and p-type doped silicon

Photovoltaic cells employ semiconductors. Elementary semiconductors are elements from group IV, such as silicon (Si) or germanium (Ge). If atoms from group V, such as phosphorus (P) and antimony (Sb), are planted into a silicon crystal lattice, the fifth electron cannot participate in the binding with the surrounding Si atoms and hence is donated to the conduction band. The embedding of group V atoms is called n-doping. The impurity atoms are referred to as donors. On the other hand, if silicon is doped with atoms from group III, such as boron (B) or aluminum (Al), an electron is missing and a hole is generated. Very little energy is needed to move this hole around (actually, electrons also move, but the holes are the majority charge carriers in this case). In this situation, the silicon is called p-doped and impurity atoms are called acceptors. The crystal structure of n-doped and p-doped silicon is shown in Fig. 1.5 [7].

When p-type and n-type semiconductors are placed in contact, a so-called p–n junction is formed. Electrons will flow from the n-type side to the p-type side, with

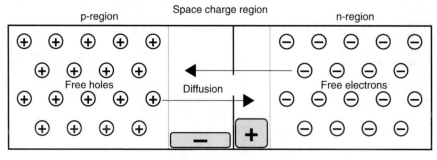

Fig. 1.6 A space charge region is formed by diffusion of electrons and holes at the p–n junction

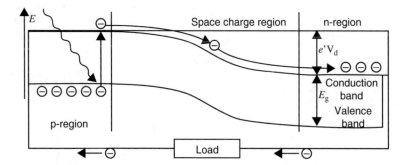

Fig. 1.7 Solar cell principles with the energy band model. E/e energy, E_g bandgap value, V_d diffusion voltage

holes flowing in the other direction, until an equilibrium state is reached (Fig. 1.6). This will create a space charge region with an electric field blocking further transfer of charge carriers. This is how band bending occurs, as shown in Fig. 1.7 [1].

1.6 The Photovoltaic Market

The rapid growth of the photovoltaic market began in the 1980s as a result of application of multimegawatt photovoltaic plants for power generation. The present photovoltaic market has grown at very high rates (30–40%), similar to those of the telecommunications and computer sectors. By 2009, world photovoltaic production had increased to 10.66 GW (Fig. 1.8) [8]. This became possible because of technology cost reductions and market development, reflecting increasing awareness of the versatility, reliability, and economy of photovoltaic electricity supply systems. The major market segments served by this industry comprise consumer applications, remote industrial systems, developing countries, and grid-connected systems.

Fig. 1.8 Evolution of the world cumulative installed photovoltaic capacity up to 2050. Historical data are denoted by *black and red circles*; forecast data are denoted by *black and white circles*, *green and white squares*, and *blue and white diamonds*

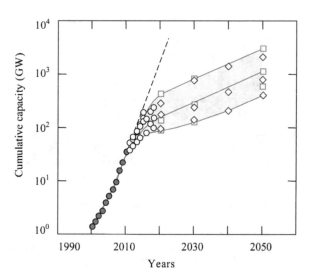

Table 1.1 Top ten photovoltaic cell/module producers worldwide

Producer	Capacity (MW)
First Solar	1011
Suntech Power	704
Sharp	595
Q-Cells	537
Yingly Green Energy	525
JA Solar	509
Kyosera	400
Trina Solar	399
Sunpower	398
Gintech	368

Of particular interest is the strong differential growth rate in rural applications, which now account for nearly half of the total photovoltaic market. The second largest market is industrial applications. Photovoltaic applications are progressively finding markets mainly in the USA, Japan, the European Union (mostly Germany), and China/Taiwan [9].

In 2009 the power production figures for photovoltaic cells and modules in the USA, Japan, the European Union, and China/Taiwan were 595 MW, 1.5 GW, 1.93 GW, and 5.19 GW, respectively. Recently, China/Taiwan became the leading photovoltaic producers, and world photovoltaic production has almost doubled. Worldwide producers are listed and presented in Table 1.1. According to Razykov et al. [8], another 6.9 GW of power cells and 6.7 GW of module capacity are being added, mostly in China/Taiwan and Japan, bringing the total global cell and module capacities to 25.1 GW and 22.7 GW, respectively. Because of an anticipated Si shortage, this goal cannot be achieved using wafer Si, despite high annual production (30,000 US tons per year) of semiconductor-purity Si. Rapid penetration of

Table 1.2 World
photovoltaic cell production,
by technology, in 2009

Technology	Capacity (MW)
Crystalline silicon	8678
Cadmium telluride	1019
Copper indium gallium selenide	166
Amorphous silicon	796

second-generation photovoltaic cells (thin-film solar cells (TFSCs)) into the world
photovoltaic market is therefore required.

The top ten producers of photovoltaic cells and modules are listed in Table 1.2.
These companies produced a combined total of 4.92 GW in 2009, amounting to
almost 50% of world production. In terms of technology, crystalline silicon (c-Si)
and polycrystalline silicon (pc-Si) wafers are the main materials used in the world
photovoltaic industry (Table 1.2) [8]; at present, >80% of the world photovoltaic
industry is based on c-Si and pc-Si wafer technologies. CdTe technology is growing
sufficiently fast, while thin-film CIGS and amorphous silicon (a-Si)–based
photovoltaic production is still in the beginning stages, despite the remarkable
results of research and development many years ago. This lag may be due to
difficulties between laboratory and large-scale production technologies. Several
new multimegawatt thin-film plants are ready for production of these types of solar
cells, and their contribution to the world photovoltaic market may be significantly
expanded soon. It is expected that thin-film photovoltaic cell (TFPC) technologies
will play a major role in the world photovoltaic market in the near future.

1.7   Thin-Film Technology

Before we get into the detail of thin-film technology, we should first brief ourselves
on what a solar cell is all about. A solar cell, also known as a photovoltaic cell, is a
solid-state device that converts energy from sunlight into direct current (DC). These
cells are assembled in modules, also known as solar panels. The greatest efficiency
observed with Si to date was 24.2%, achieved by Sunpower, a photovoltaic company
based in San Jose (CA, USA). Although there is a lot of academic research going on
around the world, with laboratory efficiencies reaching 40.7% being reported by the
Boeing subsidiary Spectrolab, efficiency values at the industry scale are still limited
to a market average of 12–18% [10].

Figure 1.9 shows the layer-by-layer deposition of a general thin-film polycrystalline solar cell. Layers of different materials are deposited one by one onto a substrate,
so we start off with the substrate, which can be glass, silicon, etc. Next comes the front
contact, which is generally molybdenum, after which come the actual semiconductor
layers: first the p layer and then the n layer. The junction between the p and n layers is
shown in Fig. 1.9. Over them comes the antireflection coating, which consists of a thin
dielectric layer of a specific thickness. The coating helps to prevent sunrays that hit the
cell from being reflected back (Fig. 1.10). The interference effects in the coating cause

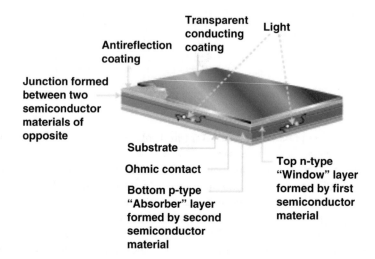

Fig. 1.9 Typical polycrystalline thin-film solar cell, which represents each layer deposited to form the final cell

the wave reflected from the antireflection coating top surface to be out of phase with the wave reflected from the semiconductor surface. These out-of-phase reflected waves destructively interfere with one another, resulting in zero net reflected energy [11]. The final layer over the antireflective coating is the top contact. Typical metals used here are aluminum, gold, silver, etc. Figure 1.10 gives a better understanding of how the antireflection coating works [11].

Thin-film solar cells (TFSCs), also known as thin-film photovoltaic cells (TFPCs), are cells made by deposition of one or more thin layers of photovoltaic material onto a substrate. The thickness of these cells varies from nanometers to the micrometer range. Numerous photovoltaic materials are used in fabrication of thin films, and different substrates are used, such as silicon, glass, and certain metals, including aluminum. The most commonly used substrates are wafers of single crystalline and polycrystalline silicon. Ingots are grown using the Czochralski method or by control of solidification of silicon in a crucible or mold. These ingots are then sliced into thin wafers, with a thickness >150 μm, by use of fine saws (Fig. 1.11). These wafers are then polished and processed until they are ready to be shipped to solar cell manufacturers. The high material and processing costs make it difficult for the price of photovoltaic technology to be reduced below US$0.10 kW/h.

The advantage that thin films have over the silicon-based approach are better throughput and lower material and labor costs. Because extremely thin layers are used, adhesion to the substrate is a major concern. A few materials, such as zinc, do not stick to the glass when deposited by CVD. (This issue is discussed in greater detail in later chapters of this volume.) Another problem with thin films is lack of uniformity of deposition, causing the thickness to vary across even small surface areas of the substrates.

At a laboratory level, good-quality cells can be produced with uniform thickness and composition, producing good efficiency rates. These are generally manufactured on

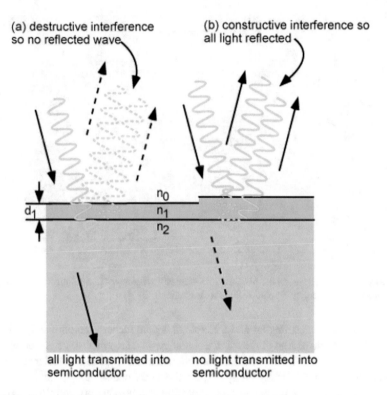

(a) destructive interference so no reflected wave

(b) constructive interference so all light reflected

n_0

d_1

n_1

n_2

all light transmitted into semiconductor

no light transmitted into semiconductor

Fig. 1.10 Constructive and destructive interference of light rays hitting an antireflective coating layer of a solar cell

Fig. 1.11 Typical silicon ingots of various wafer sizes, grown using the Czochralski method

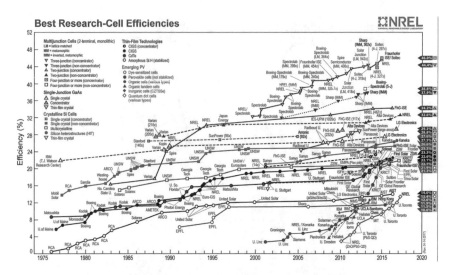

Fig. 1.12 Developments on the best research solar cell efficiencies in multijunction, crystalline, thin-film and other emerging photovoltaic technologies

small substrates measuring around a couple of square inches (approximately 13 cm²). However, implementation of the same technology on a larger substrate is a completely different ball game. All of the factors mentioned above lead to macroscopic defects, which affect the yield and reliability of the cell over the course of its lifetime.

On the basis of the materials used, thin films can be categorized into amorphous silicon (a-Si), cadmium telluride (CdTe), copper indium gallium selenide (CIS or CIGS), or dye-sensitized solar cells (DSCs). Figure 1.12 shows the various solar cell manufacturing technologies and where they stand as far as efficiency is concerned. We can see that thin-film technology is still evolving in comparison with the well-established silicon industry, which is far ahead, thanks to the huge semiconductor industry, which accelerated the groundwork [12].

Nowadays, a vigorous search is on for new alternative absorber TFSCs made from elements that are earth abundant, sustainable, affordable, and nontoxic [13]. In this quest, Cu_2ZnSnS_4 (CZTS) and $Cu_2ZnSn(S,Se)_4$ (CZTSSe) have emerged as the most promising materials. However, there is a simpler material, copper tin sulfide (Cu_2SnS_3; CTS), which has received very little attention. CTS is a p-type semiconductor with a direct bandgap of 1.1 eV and an absorption coefficient around 10^5 cm^{-1}. About 25 years ago, Kuku and Fakolujo [14] demonstrated, for the first time, use of vacuum-evaporated CTS thin films for photovoltaic devices. However, recent attempts [15, 16] to prepare CTS films resulted in films with CuS_x as an impurity.

1.8   Research Objectives

1. To explore use of copper tin sulfide and a prospective low-cost absorber layer of thin-film solar cells

2. To investigate the potential device structure of copper tin sulfide–based thin-film solar cells
3. To optimize the cell structure in order to achieve higher conversion efficiency of copper tin sulfide–based solar cells

1.9   The Structure of This Book

Research has been conducted on the basis of the aforementioned scope and objectives. Figure 1.13 shows the structure of this book.

CHAPTER 1

INTRODUCTION

- Solar photovoltaic cells and other alternatives
- PRINCIPLES OF OPERATIONS OF SOLAR
- Problem statement and research objectives

CHAPTER 2

LITERATURE REVIEW

- Solar energy and cells and p-n junction
- Various types of solar cells and their efficiencies
- Silicon solar cell and light trapping method

CHAPTER 3

METHODOLOGY

- Numerical modeling computer program with SCAPS
- Cell structure

CHAPTER 4

RESULTS AND DISCUSSION

- Numerical Modeling Results

CHAPTER 5

CONCLUSION

- Research Summary & Future Prospects

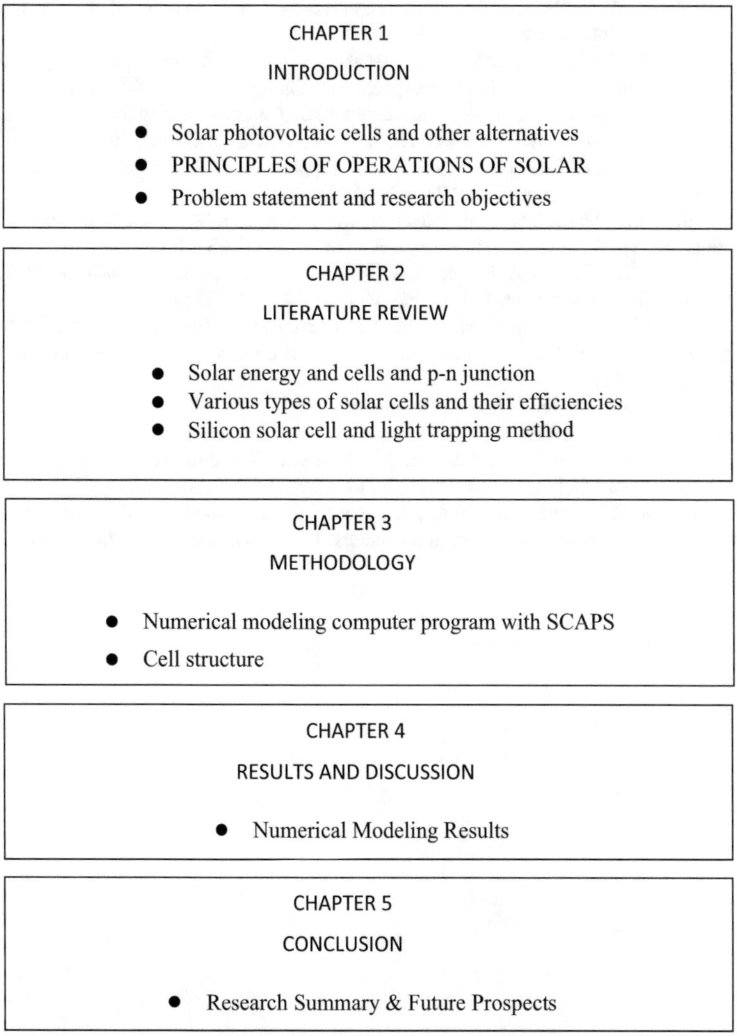

Fig. 1.13 General overview of this book

References

1. V. Quaschning, *Understanding renewable energy systems* (Earthscan/James & James, London, 2005)
2. A. Smee, *Elements of electro-biology: or the voltaic mechanism of man; of electro-pathology, especially of the nervous system; and of electro-therapeutics* (Longman, Brown, Green & Longmans, London, 1849)
3. J.L. Sawin, *Mainstreaming renewable energy in the 21st century* (Worldwatch Institute, Washington, DC, 2004)
4. P. Moriarty, D. Honnery, Hydrogen's role in an uncertain energy future. Int. J. Hydrog. Energy **34**, 31–39 (2009)
5. J.R. Ritter, The "hot issue" market of 1980. J. Bus. **57**, 215–240 (1984)
6. R.H. Wiser, Renewable energy finance and project ownership: the impact of alternative development structures on the cost of wind power. Energy Policy **25**, 15–27 (1997)
7. B. O'Regan, M. Grfitzeli, A low-cost, high-efficiency solar cell based on dye-sensitized colloidal TiO_2 film. Nature **353**, 24 (1991)
8. T. Razykov, C. Ferekides, D. Morel, E. Stefanakos, H. Ullal, H. Upadhyaya, Solar photovoltaic electricity: current status and future prospects. Sol. Energy **85**, 1580–1608 (2011)
9. S. Hegedus, A. Luque, Achievements and challenges of solar electricity from photovoltaics, in *Handbook of photovoltaic science and engineering*, (Wiley, Chichester, 2011), pp. 1–38
10. A. Shah, P. Torres, R. Tscharner, N. Wyrsch, H. Keppner, Photovoltaic technology: the case for thin-film solar cells. Science **285**, 692–698 (1999)
11. J.A. Hiller, J.D. Mendelsohn, M.F. Rubner, Reversibly erasable nanoporous anti-reflection coatings from polyelectrolyte multilayers. Nat. Mater. **1**, 59–63 (2002)
12. D. Macdonald, A. Cuevas, A. Kinomura, Y. Nakano, L. Geerligs, Transition-metal profiles in a multicrystalline silicon ingot. J. Appl. Phys. **97**, 033523–033527 (2005)
13. L. Han, A. Islam, H. Chen, C. Malapaka, B. Chiranjeevi, S. Zhang, X. Yang, M. Yanagida, High-efficiency dye-sensitized solar cell with a novel co-adsorbent. Energy Environ. Sci. **5**, 6057–6060 (2012)
14. T.A. Kuku, O.A. Fakolujo, Photovoltaic characteristics of thin films of Cu_2SnS_3. Sol. Energy Mater. **16**, 199–204 (1987)
15. M. Bouaziz, J. Ouerfelli, M. Amlouk, S. Belgacem, Structural and optical properties of Cu_3SnS_4 sprayed thin films. Phys. Status Solidi A **204**, 3354–3360 (2007)
16. P. Fernandes, P. Salomé, A.F. Cunha, A study of ternary Cu_2SnS_3 and Cu_3SnS_4 thin films prepared by sulfurizing stacked metal precursors. J. Phys. D. Appl. Phys. **43**, 215403 (2010)

Chapter 2
Solar Energy-Based Semiconductors: Working Functions and Mechanisms

2.1 Introduction of Solar Energy

The energy of solar radiation is directly utilized in mainly two forms:

1. Direct conversion into electricity that takes place in semiconductor devices called solar cells
2. Accumulation of heat in solar collectors.

The direct conversion of solar radiation into electricity is often described as a photovoltaic (PV) energy conversion because it is based on the photovoltaic effect. In general, the photovoltaic effect means the generation of a potential difference at the junction of two different materials in response to visible or other radiation. The whole field of solar energy conversion into electricity is therefore denoted as the "photovoltaic." Photovoltaic literally means "light-electricity," because "photo" is a stem from the Greek word "phõs" meaning light and "Volt" is an abbreviation of Alessandro Volta's (1745–1827) name who was a pioneer in the study of electricity. Since a layman often does not know the meaning of the word photovoltaic, a popular and common term to refer to PV solar energy is solar electricity.

Among the radical ways to reduce the cost of solar modules and to increase drastically the volume of their production is the transition to thin-film technology, the use of direct-gap semiconductors deposited on a cheap large-area substrate (glass, metal foil, plastic). We start with the fact that the direct-gap semiconductor can absorb solar radiation with a thickness, which is much smaller than the thickness of the silicon wafer. This is illustrated by the results of calculations in Fig. 2.1 similar to those performed for the single-crystalline silicon. Calculations were carried out for direct-gap semiconductors, which is already used as absorber layers of solar modules: a-Si, CdTe, $CuInSe_2$, and $CuGaSe_2$. As expected, the absorptivity of solar radiation of direct-gap semiconductors in general is much stronger compared to

© The Author(s), under exclusive license to Springer Nature Switzerland AG 2019
I. S. Amiri, M. Ariannejad, *Introducing CTS (Copper-Tin-Sulphide) as a Solar Cell by Using Solar Cell Capacitance Simulator (SCAPS)*, SpringerBriefs in Electrical and Computer Engineering, https://doi.org/10.1007/978-3-030-17395-1_2

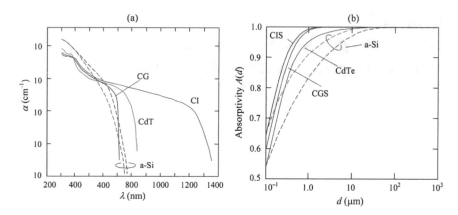

Fig. 2.1 Absorption curves (**a**) and dependence of absorptivity solar radiation in the $h\upsilon \geq E_g$ spectral range on the absorber layer thickness d (**b**) for amorphous silicon (a-Si), cadmium telluride (CdTe), copper-indium diselenide (CIS), and copper gallium diselenide (CGS)

crystalline silicon but the curves noticeably differ among themselves (in the references, the absorption curves for a-Si are somewhat different). The complete absorption of solar radiation by amorphous silicon n (a-Si) in the $\lambda \leq \lambda_g = hc/Eg$. Spectral range is observed at its thickness $d > 30$–60 µm, and 95% of the radiation is absorbed at a thickness of 2–6 µm (Fig. 2.1a). These data are inconsistent with the popular belief that in a-Si, as a direct-gap semiconductor, the total absorption of solar radiation occurs at a layer thickness of several microns. The total absorption of solar radiation in CdTe occurs if the thickness of the layer exceeds 20–30 µm, and 95% of the radiation is absorbed if the layer is thinner than ~1 µm. Absorptivity of the CuInSe$_2$ and CuGaSe is even higher. The complete absorption of radiation in these materials takes place at a layer thickness of 3–4 µm, and 95% of the radiation is absorbed if the thickness of layer is only 0.4–0.5 µm.

In fact, to collect photo-generated charge carriers, it is necessary to have a diffusion length of minority carriers in excess of the thickness of the absorbing layer. In the case of crystalline Si, the photo-generated carriers must be collected at a thickness of 1–200 microns and 2 orders of magnitude smaller than in the case of CdTe, CIS, or CGS. From this it follows that in the solar cell based on direct-gap semiconductor, the diffusion length L may be about two. In fact, to collect photo-generated charge carriers, it is necessary to have a diffusion length of minority carriers in excess of the thickness of the absorbing layer. In the case of crystalline Si, the photo-generated carriers must be collected at a thickness of 1–200 microns and 2 orders of magnitude smaller than in the case of CdTe, CIS, or CGS. From this it follows that in the solar cell based on direct-gap semiconductor, the diffusion length L may be about two orders of magnitude smaller, i.e., the carrier lifetime can be by 4 orders shorter ($L \sim \tau^{1/2}$) [1].

Thus, the manufacture of thin-film solar modules based on the direct-gap semiconductors does not require costly high purification and crystallinity of the material as it is needed in the production of modules based on crystalline, multicrystalline, or ribbon silicon. Thin-film technology has a number of other significant merits. While Si devices are manufactured from wafers or ribbons and then processed and assembled to form a module, in thin-film technology many cells are simultaneously made and formed as a module. The layers of solar cells are deposited sequentially on moving substrates in a continuous highly automated production line (conveyor system) and, importantly, at temperatures not exceeding 200–650 °C compared with 800–1450 °C for the main processes of c-Si. This minimizes handling and facilitates automation leading to the so-called monolithic integration.

Thin-film solar modules offer the lowest manufacturing costs, and are becoming more prevalent in the industry because they allow to improve manufacturability of the production at significantly larger scales than for wafer or ribbon Si modules. Therefore, it is generally recognized that the contribution of thin-film technology in solar energy will be to grow from year to year faster. Many analysts believe that it is only a matter of time before thin films would replace silicon wafer-based solar cells as the dominant photovoltaic technology. Unquestionable leaders in thin-film technologies are solar cells on amorphous silicon (a-Si), copper-indium-gallium diselenide ($CuIn_xGa_{1-x}Se_2$), and cadmium telluride (CdTe), whose market share is expanding every year [2]. The rest of the thin-film technologies are yet too imma-ture to appear in the market but some of them are already reaching the level of industrial production. Below these technologies will first be briefly described, and a more detailed analysis of solar modules based on a-Si, CdTe, and CIGS is allocated in separate subsections.

2.2   Recombination in Solar Cells

Electricity generation by Si solar cells relies on collection of photo-generated carri-ers at the p–n junction. Specifically, the photo-generated minority carriers at each side of the cell (electrons generated in the p-side and holes generated in the n-side) should travel to the junction, where they are swept to the other side of the electric field at the junction. Any loss or recombination of these excited carriers, before reaching the junction, contributes to a loss in cell performance [3]. In a conventional p-type Si solar cell shown in Fig. 2.2, recombination can occur in five regions:

(a) Front surface
(b) n-doped region or the emitter
(c) Depletion region of the p–n junction
(d) p-doped region or the base
(e) Back surface.

There are three fundamental recombination mechanisms in a bulk semiconductor:

(a) Radiative or band-to-band recombination
(b) Auger recombination

Fig. 2.2 Five common recombination sites in a p-type Si solar cell

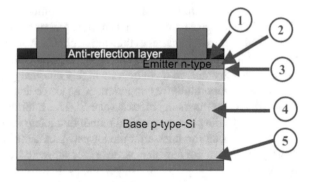

(c) Recombination through defect or trap levels (often referred to as Shockley–Read–Hall recombination [SRH] recombination). Prior to going into the mathematical expressions of each of the three recombination mechanisms, it is instructive to introduce two main quantities that are commonly used to assess the volume recombination activity:

- Volume recombination rate, U_V (cm^{-3} s^{-1}); a recombination rate of carriers per unit volume per unit time.
- Recombination lifetime, t(s); an average time that excess carriers can survive before recombining the lifetime can be expressed in terms of the volume recombination rate and the excess carrier concentration (Δn) as:

$$\tau \equiv \frac{\Delta n}{U} \tag{2.1}$$

- Note that the use of the lifetime definition is valid only when there are excess carriers in the volume (nonthermal equilibrium). The excess carrier concentration or the injection level has a significant impact on the recombination behavior [3]. Generally, the injection level is defined with respect to the doping concentration as follows:
  - Low-level injection (LLI) corresponds to the condition where the excess carrier concentration is smaller than the doping concentration by at least an order of magnitude.
  - High-level injection (HLI) corresponds to the condition where the excess carrier concentration is larger than the doping concentration by at least an order of magnitude.

Radiative recombination (also referred to as band-to-band recombination) corresponds to the recombination process where a free electron falls directly from the conduction band and recombines with a free hole in the valence band with all or most of the excess energy dissipated in the form of a photon (see also Fig. 2.3).

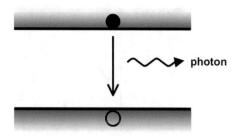

Fig. 2.3 Schematic representation of radiative recombination

$$U_{BB = Bnp} \tag{2.2}$$

where B is the coefficient of radiative recombination, and is the free-electron concentration and p is the free-hole concentration. From detailed balance calculation, the value of B for Si was calculated to be $2 \times 10^{-15} \mathrm{cm^3\ s^{-1}}$ [4]. The experimental value, however, was found to be much higher, $9.5 \times 10^{-15} \mathrm{cm^3\ s^{-1}}$ [5]. At thermal equilibrium ($\Delta n = 0$), U_{BB} is equivalent to the thermal generation rate, G_{th}; the expression for U_{BB} at thermal equilibrium is

$$U_{BB} = G_{th} = Bn_0 p_0 = Bn_i^2 \tag{2.3}$$

where n_0 and p_0 are thermal equilibrium concentrations of free electrons and free holes and n_i is the intrinsic carrier concentration. From Eqs. (2.1) and (2.2), the radiative recombination lifetime can readily be obtained as

$$\tau_{BB} = \frac{\Delta n}{B.(n_0 + \Delta n).(p_0 + \Delta n)} \tag{2.4}$$

Consequently, the expression for the radiative recombination lifetime under low and high injection is as follows:

$$\tau_{BB,lli} = \frac{1}{B.N_{doped}} \quad \text{and} \quad \tau_{BB,hli} = \frac{1}{B.\Delta n} \tag{2.5}$$

where N_{doped} is the donor (N_D) or the acceptor (N_A) concentration for n- or p-type semiconductors, respectively. Note that the radiative recombination lifetime stays constant at low injection and decreases with injection in intermediate and high injection regimes. The radiative recombination in an indirect band gap semiconductor such as Si is considered to be small compared to other types of recombination. This is because the process involves phonon as the fourth particle (apart from an electron, a hole, and a photon) to conserve the momentum (see Fig. 2.4).

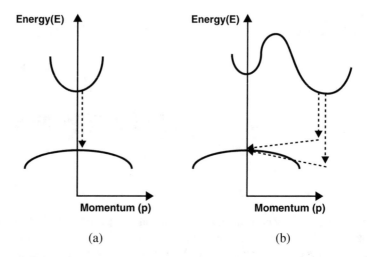

Fig. 2.4 Schematic representation of radiative recombination for (**a**) direct band gap (e.g., GaAs) and (**b**) indirect band gap (e.g., Si)

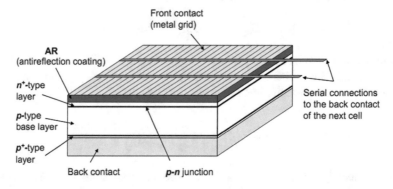

Fig. 2.5 A typical structure of a c-Si solar cell

In most of today solar cells the absorption of photons, which results in the generation of the charge carriers, and the subsequent separation of the photo-generated charge carriers take place in semiconductor materials. Therefore, the semiconductor layers are the most important parts of a solar cell; they form the heart of the solar cell. There are a number of different semiconductor materials that are suitable for the conversion of energy of photons into electrical energy, each having advantages and drawbacks. In this chapter the most important semiconductor properties that determine the solar cell performance will be discussed.

The crystalline silicon (c-Si) solar cell, which dominates the PV market at present, has a simple structure, and provides a good example of a typical solar cell structure. Figure 2.5 shows the essential features of c-Si solar cells. An absorber material is typically a moderately doped p-type square wafer having thickness around 300 μm and an area of 10×10 cm^2 or 12.5×12.5 cm^2 on both sides of the c-Si wafer a highly doped layer is formed, n$^+$-type on the top side and p$^+$-type on the back side, respectively. These highly doped layers help to separate the photo-generated charge carriers from the bulk of the c-Si wafer. The trend in the photovoltaic industry is to reduce the thickness of wafers up to 250 μm and to increase the area to 20×20 cm^2 [6].

In addition to semiconductor layers, solar cells consist of a top and bottom metallic grid or another electrical contact that collects the separated charge carriers and connects the cell to a load. Usually, a thin layer that serves as an antireflective coating covers the top side of the cell in order to decrease the reflection of light from the cell. In order to protect the cell against the effects of outer environment during its operation, a glass sheet or other type of transparent encapsulant is attached to both sides of the cell. In case of thin-film solar cells, layers that constitute the cell are deposited on a substrate carrier. When the processing temperature during the deposition of the layers is low, a wide range of low-cost substrates such as glass sheet, metal, or polymer foil can be used.

1. Concentrations of doping atoms, which can be of two different types: donor atoms which donate free electrons, N_D, or acceptor atoms, which accept electrons, N_A. The concentrations determine the width of a space-charge region of a junction.
2. Mobility, μ, and diffusion coefficient, D, of charge carriers that characterize carriers transport due to drift and diffusion, respectively.
3. Lifetime, τ, and diffusion length, L, of the excess carriers that characterize the recombination–generation processes.
4. Band gap energy, E_G, absorption coefficient, α, and refractive index, n, that characterize the ability of a semiconductor to absorb visible and other radiation.

2.3   Semiconductor Properties

Semiconductors are a group of materials having electrical conductivities intermediate between metal and insulator. The conductivity of these materials can be varied by changes in temperature, optical excitation, and impurity content, this variability of electrical properties makes the semiconductor material natural choices for electronic device investigation. Semiconductor materials are found in column IV and neighboring column of the periodic table. The column IV semiconductor is called an elemental semiconductor because they are composed of single species of atom. In addition to elemental material, compounds of column III and V also can make up compound semiconductor. The elemental semiconductor Ge was widely used in the early days of semiconductor development for transistors and diodes. Silicon is now used majority in transistors, rectifiers, and integrated circuits. One of the most important characteristic of a semiconductor which distinguishes it from metals and insulators is its energy band gap. This property determines the wavelength of the light that can be absorbed or emitted by the semiconductor.

2.3.1   Atomic Structure

The atomic number of Si atom is 14; it means there are 14 electrons orbiting the nucleus. In the ground state configuration a Si atom has four valence electrons. These valence electrons are most important because they form the bonds with other Si atoms. Two Si atoms are bonded together when they share each other's valence electron. This is the so-called covalent bond that is formed by two electrons. Since the Si atom has four valence electrons it can be covalently bonded to four other Si atoms. In the crystalline form each Si atom is covalently bonded to four neighboring Si atoms. All bonds have the same length and the angles between the bonds are equal. The number of bonds that an atom has with its immediate neighbors in the atomic structure is called the coordination number or coordination. Thus, in single crystal silicon, the coordination number for all Si atoms is four, we can also say that Si atoms are fourfold coordinated. A unit cell can be defined, from which the crystal lattice can be reproduced by duplicating the unit cell and stacking the duplicates next to each other. Such a regular atomic arrangement is described as a structure with long range order. A diamond lattice unit cell represents the real lattice structure of single crystal silicon. Figure 2.6a shows the arrangement of the unit cell and Fig. 2.6b the atomic structure of single crystal silicon. One can determine from Figure 2.6a that there are effective eight Si atoms in the volume of the unit cell. When a lattice constant of c-Si is 5.4 Å one can easily calculate that there are approximately 5×10^{22} Si atoms per cm^3. Figure 2.6 shows the crystalline Si atomic structure with no foreign atoms. In practice, a semiconductor sample always contains some impurity atoms. When the concentration of impurity atoms in a semiconductor is insignificant we refer to such semiconductor as an intrinsic semiconductor.

At practical operational conditions, for example, room temperature, there are always some of the covalent bonds broken. The breaking of the bonds results in liberating the valence electrons from the bonds and making them mobile through

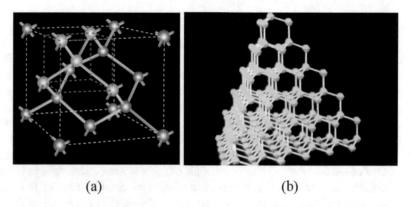

(a)   (b)

Fig. 2.6 (a) A diamond lattice unit cell represents a unit cell of single crystal Si, (b) the atomic structure of a part of single crystal Si

the crystal lattice. We refer to these electrons as free electrons (henceforth simply referred as electrons). The position of a missing electron in a bond, which can be regarded as positively charged, is referred to as a hole [6]. This situation can be easily visualized by using the bonding model shown in Fig. 2.6.

In the bonding model the atomic cores (atoms without valence electrons) are represented by circles and the valence or bonding electrons are represented by lines interconnecting the circles. In case of c-Si one Si atom has four valence electrons and four nearest neighbors. Each of the valence electron is equally shared with the nearest neighbor. There are therefore eight lines terminating on each circle. In an ideal Si crystal at 0 °K all valence electrons take part in forming covalent bonds between Si atoms and therefore are no free electrons in the lattice. This situation is schematically shown in Fig. 2.7a. At temperatures higher than 0 °K the bonds start to break due to absorbing thermal energy. This process results in the creation of mobile electrons and holes. Figure 2.7b showed a situation when a covalent bond is broken and one electron departs from the bond leaving a hole behind. A single line between the atoms in Fig. 2.7b represents the remaining electron of the broken bond. When a bond is broken and a hole created, a valence electron from a neighboring bond can "jump" into this empty position and restore the bond. The consequence of this transfer is that at the same time the jumping electron creates an empty position in its original bond. The subsequent "jumps" of a valence electron can be viewed as a motion of the empty position, hole, in the direction opposite to the motion of the valence electron through the bonds.

Because breaking of a covalent bond leads to the formation of an electron–hole pair, in intrinsic semiconductors the concentration of electrons is equal to the

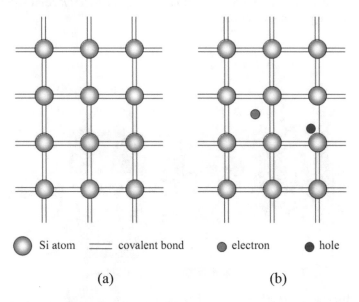

Si atom ══ covalent bond ● electron ● hole

(a) (b)

Fig. 2.7 The bonding model for c-Si. (**a**) No bonds are broken. (**b**) A bond between two Si atoms is broken resulting in a mobile electron and hole

concentration of holes. At 300 °K there are approximately 1.5×10^{10} broken bonds per cm^3 in the intrinsic c-Si. This number then gives also the concentration of holes, p, and electrons, in the intrinsic c-Si. It means that at the temperature of 300 °K the electron and hole is $n = p = 1.5 \times 10^{10}$ cm^{-3}. This concentration is called the intrinsic carrier concentration and is denoted n_i.

The complete structure of solar cells consists of a semiconductor, P–N junction photodiodes with a large light-sensitive area. In order to provide better understanding about solar cell structure, brief definitions about these solar components will be provided before the overall concept. Firstly, a semiconductor is a class of materials whose electrical properties lie between those of conductors (metals) and insulators (non-metals). At the atomic level, semiconductors are crystals that in their pure state are resistive, but when the proper impurities are added—this process is called doping, semiconductors display much lower resistance along with other interesting and useful properties [7]. Depending on the selection of impurities added, two types of semiconductor could be created: N type (electron-rich) or P type (electron-poor). Examples for semiconductors are silicon and germanium.

P–N junction is the basic formation of P type and N type semiconductors by intimate contact. The purpose of this formation conducts electric current with one polarity of applied voltage (forward bias) without conducting the opposite polarity (reverse bias). Finally, photodiodes are basically P–N junctions, which are specifically designed to optimize their inherent photosensitivity as shown in Fig. 2.8. Together, these materials create three main energy-conversion layers of a solar cell. The first layer necessary for energy conversion is the top junction layer, which is made of N-type semiconductor. The next layer is the absorber layer called the P–N junction. The last of the energy-conversion layers is the back junction layer, which is made of P-type semiconductor [8].

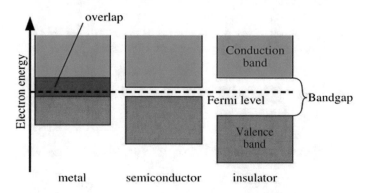

Fig. 2.8 The band gap theory

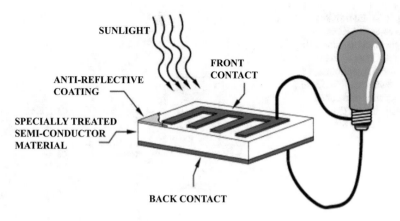

Fig. 2.9 Solar cell circuit

2.3.2 Mechanism of Solar Cells

The mechanism of solar cells is quite simple. Basically, when light energy strikes a solar cell [9], electrons are knocked loose from the atoms in the semiconductor material. If electrical conductors are attached to the positive and negative sides, which form an electrical circuit, the electrons can be captured in the form of an electric current, that is, electricity. This electricity can be used to power a load [10]. Figure 2.9 represents the mechanism of a solar cell.

2.4 P–N Junction Solar Cells

When photons strike a solar cell, pieces of the photons are reflected back. Only those photons can be absorbed and generate electron–hole pairs which their energies are above E_g. As it is shown in Fig. 2.10, those photons with the energy below E_g pass through the device without any contribution to the resulting, as marked by process 1. Since the photons absorbed by the solar cells, prior to extraction of the photo-generated electron–hole pairs to the load, the main energy loss occurs in process 2, i.e., the thermalization loss, which is due to the rapid thermal relaxation for the electron–hole pairs activated by high-energy (E) photons, in the form of releasing the energy of $(E–E_g)$ to generate photons. The other energy loss mechanisms include junction and contact voltage losses (3 and 4), and recombination loss 5. Both processes 1 and 2 account for ~50% of the solar energy loss and are related to the fixed band gap of the semiconductor (accordingly photons with too little or too much energy cannot be effectively utilized) [11].

Fig. 2.10 Energy loss processes in a standard single junction solar cell: (1) non-absorption of sub-band gap photons; (2) thermalization loss; (3) and (4) junction and contact voltage losses; (5) recombination loss

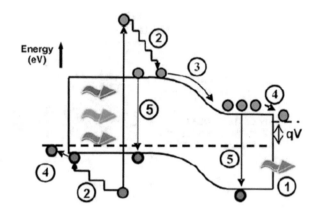

There are various photovoltaic cells having different materials, structures, and fabrication methods. Table 2.1 shows the record efficiency of single cells and sub-modules, respectively [12, 13]. Single junction solar cells with crystalline silicon are the dominant product in photovoltaic markets and 80–90%. All solar cell modules installed today are wafer-based Si ones [14]. As mentioned above, low cost and high efficiency cells with a large area are the main challenges in developing and manufacturing solar cells. At the time of writing silicon is the optimum material in view of material cost since it is the second most abundant element among the natural non-renewable resources currently available. However, expensive equipment for growing the silicon ingot, high temperatures needed for thermal treatment like diffusion or annealing, and toxic chemicals to treat the surface increases production costs for silicon cells. The price reductions in this technology are close to saturation.

2.5 Copper-Tin Sulfide Solar Cell

In recent years, there has been a great deal of interest in the study of nontoxic semiconductors from both the fundamental and technological points of view for solar cell materials. Ternary semiconductors such as copper-tin sulfides (CTS) have attracted a great deal of attention as excellent absorber materials because of their high absorption coefficient ($>10^4$ cm^{-1}) for photovoltaic cells, and are a suitable candidate for nonlinear optical materials. Furthermore, they have outstanding optical–thermal–mechanical properties as small or mid-band gap semiconductors [36, 37].

2.5.1 Copper-Tin Sulfides Solar Cell: Overview

Many semiconductors of CTS phases have been reported, such as Cu_4SnS_4, Cu_2SnS_3, $Cu_4Sn_7S_{16}$, $CuSn_3$, and Cu_3SnS_4 [38]. Among these ternary sulfides, Cu_2SnS_3 and Cu_3SnS_4 are the most promising compounds, and these can be used in solar cells

Table 2.1 Confirmed terrestrial cells and submodule efficiencies measured under the global AM 1.5 spectrum (1000 W/m^2) at 25 °C

Classification	Effic. (%)	Voc (V)	Jsc (mA/cm^2)	FF (%)	Description
Silicon					
Si (crystalline)	25.0 ± 0.5	0.706	42.7	82.8	[15]
Si (multicrystalline)	20.4 ± 0.5	0.664	38.0	80.9	[16]
Si (thin-film transfer)	16.7 ± 0.4	0.645	33.0	78.2	[17]
Si (thin-film submodule)	10.5 ± 0.3	0.492	29.7	72.1	[18]
III–V cells					
GaAs (thin film)	27.6 ± 0.8	1.107	29.6	84.1	[19]
GaAs (multicrystalline)	18.4 ± 0.5	0.994	23.2	79.7	[20]
InP (crystalline)	22.1 ± 0.7	0.878	29.5	85.4	[21]
Thin-film chalcogenide					
CIGS (cell)	19.6 ± 0.6	0.713	34.8	79.2	[22, 23]
CIGS (submodule)	16.7 ± 0.4	0.661	33.6	75.1	[24]
CdTe (cell)	16.7 ± 0.5	0.845	26.1	75.5	[25]
CdTe (submodule)	12.5 ± 0.4	0.838	21.2	70.5	ASP Hangzhou, 8 serial cells
Amorphous /nanocrystalline Si					
Si (amorphous)	10.1 ± 0.3	0.886	16.75	67	[26]
Si (nanocrystalline)	10.1 ± 0.2	0.539	24.4	76.6	[27]
Photochemical					
Dye sensitized	10.4 ± 0.3	0.729	22	65.2	[28]
Dye sensitized (submodule)	9.9 ± 0.4	0.719	19.4	71.4	[29]
Organic					
Organic polymer	8.3 ± 0.3	0.816	14.46	70.2	[30]
Organic (submodule)	3.5 ± 0.3	8.62	0.847	48.3	[31]
Multijunction devices					
GaInP/GaAs/Ge	32.0 ± 1.5	2.622	14.37	85	Spectrolab (monolithic)
GaAs/CIS(thin film)	25.8 ± 1.3	–	–	–	[32]
a-Si/μc-Si (thin-film cell)	11.9 ± 0.8	1.346	12.92	68.5	[33]
a-Si/ic-Si (thin-film submodule)	11.7 ± 0.4	5.462	2.99	71.3	[34]
Organic (2-cell tandem)	8.3 ± 0.3	1.733	8.03	59.5	[35]

owing to their high absorption coefficient and their optimal direct band gap for solar energy conversion [39, 40]. A few attempts have been made to prepare these compounds. A solvothermal process to grow Cu_2SnS_3 nanocrystalline, Cu_3SnS_4 nanotubes, and nanorods with a tetragonal structure has been used by Wu et al. [41].

Characterization of Cu_2SnS_3 with a monoclinic structure and also the study of fundamental properties of tetragonal Cu_2SnS_3 and rhombohedral Cu_4Sn_7S bulk crystals have been performed by the conventional solid-state reaction [42, 43]. Using the spray pyrolysis technique, Bouaziz et al. [39, 44] have deposited and characterized cubic Cu_2SnS_3 and tetragonal Cu_3SnS_4 thin films. They also estimated an absorption coefficient of about 1.0×10^4 cm^{-1} and band gaps of 1.15 and 1.35 eV for these two layers, respectively. Fernandes et al. [38] have used sulfurization of dc magnetron sputtered Sn–Cu metallic precursors in an S_2 atmosphere at different temperatures 350, 400, and up to 520 °C for the growth of thin films with tetragonal Cu_2SnS_3, cubic Cu_2SnS_3, and orthorhombic Cu_3SnS_4 phases with high absorption coefficient close to 10^4 cm^{-1} and estimated band gap energies of about 1.35, 0.96, and 1.60 eV, respectively. Among these techniques, spray pyrolysis is a promising one, because of its simplicity and the capability for cost-effective large-area deposition, with no need of any sophisticated instrumentation. The composition, morphology, and optical and electrical properties of thin films, deposited by spray pyrolysis, can be tailored by changing the deposition parameters such as the composition of the precursor solution, the substrate temperature, the spray rate, and the nozzle-to-substrate distance. To the best of our knowledge, the physical properties of spray CTS films with triclinic structure by spray pyrolysis on a glass substrate have not been reported so far. To achieve this goal aqueous solutions of copper (II) acetate, tin chloride, and thiourea with various Sn/Cu molar ratios to assess the potential of these materials as solar cell absorber layers are used, relatively high absorption coefficient ($\sim 10^5$ cm^{-1}) and p-type conductivity.

In this study, we have investigated the corresponding samples from various viewpoints, including structural (crystallinity, composition, and surface morphology), optical (UV–vis–near-IR transmittance/reflectance spectra), and electrical resistivity properties. The crystal structure of monoclinic Cu_2SnS_3 is illustrated in Fig. 2.11. The results of Fig. 2.11 show that the structure of monoclinic Cu_2SnS_3 is based on cubic close packing of S with Cu and Sn in tetrahedral sites, and the structure is isomorphic with Cu_2GeS_3 [45] and Cu_2SiS_3 [46].

The values are acceptable as compared with those reported in the literature [45, 46]. For example, Cu-S distances of 2.265(3)–2.425(7) and 2.215(10)–2.561(8) Å for two kinds of tetrahedral Cu atoms, and Sn–S distances of 2.407(7)–2.431(8) Å for the four-coordinated Sn atoms, were observed in the Cu_4SnS_4 structure [47]. In the compounds that are isomorphic with the present subject compound, the corresponding Cu–S distances are 2.292(3)–2.373(5) Å in Cu2GeS3 [45] and 2.2835(8)–2.3875(8) Å in Cu_2SiS_3 [46].

2.5.2 TCO (Transparent Conducting Oxide)

For the further development and world-wide market growth of photovoltaic (PV) power generation, a reduction of the investment costs of the PV system is one of the major issues. One approach which promises a significant cost reduction are thin-film solar cells. An integral part of these devices are the transparent conductive oxide (TCO) layers used as a front electrode and as part of the back side reflector. When applied on the front side, TCO possesses a high transparency in the spectral region where the solar cell is operating and a high electrical conductivity. These are necessary but not sufficient properties of good TCO. For the so-called superstrate or

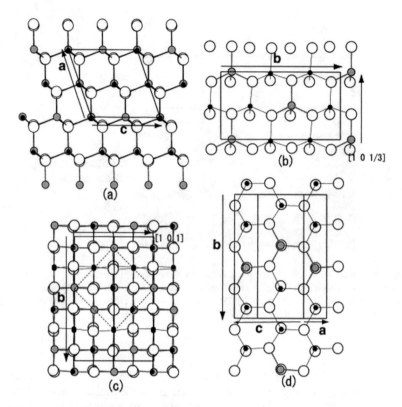

Fig. 2.11 The crystal structure of monoclinic Cu$_2$SnS$_3$. (**a**) Projection along [0 21 0], (**b**) bounded projection (0.1 <z <0.7) along [0 0 −1], (**c**) projection along [−1 0 1], and (**d**) bounded projection (0.3 < x < 0.9) along [3 0 1]. The large open, medium shadowed, and small solid circles represent S, Sn, and Cu, respectively. The thin broken lines in (**c**) represent the face-centered cubic cell of the sphalerite-type basic structure

p–i–n configuration (Fig. 2.12) where the Si layers are deposited onto a transparent substrate (e.g., glass) covered by TCO, two further conditions have to be fulfilled: strong scattering of the incoming light into the silicon absorber layer and favorable physico-chemical properties for the growth of the silicon. For example, the TCO has to be inert to hydrogen-rich plasmas or act as a good nucleation layer for growth of nano- or microcrystalline material.

For all thin-film silicon solar cells, scattering at interfaces between neighboring layers with different refractive indices and subsequent trapping of the incident light within the silicon absorber layers are crucial to gain a high efficiency (see light paths in Fig. 2.12). The reason is the absorption coefficient of amorphous and microcrystalline silicon. Whereas α is high for short wavelength light, it strongly decreases towards longer wavelengths as the light energy approaches the optical band gap of the material (about 1.8 eV for standard a-Si:H and 1.1 eV for μc-Si:H). Due to the low band gap, light in μc-Si:H can be absorbed up to the near infrared

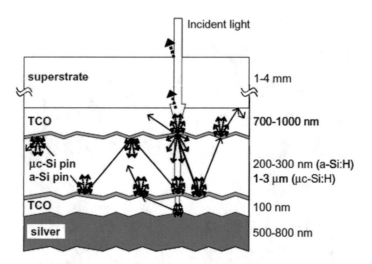

Fig. 2.12 Schematic sketch of the cross section of a silicon thin-film p–i–n solar cell (a-Si:H and/ or lc-Si:H) with rough interfaces. The thicknesses of the individual layers are typical values. The concept of light trapping is illustrated by the arrows representing incoming and scattered sunlight. Different light paths and scattering events are sketched

spectral region allowing a better utilization of the solar spectrum. However, the band gap of single-, poly-, or microcrystalline silicon is indirect thus limiting the absorption coefficient. As a result, in a thin film of not more than several micron thickness incoming light will not be completely absorbed during one single pass. On the other hand, to minimize process time and reduce light-induced degradation of amorphous Si, the absorber layer thickness should be as thin as possible. Hence, for all absorber material's optical absorption inside the silicon layers has to be enhanced by increasing the optical path of solar radiation [48].

In conclusion, TCO plays an important role in the thin-film silicon solar cell structure and has a decisive influence on the efficiencies presently achievable in state-of-the-art amorphous, microcrystalline, or "micromorph" solar cells. As the conditions for the best values of several material parameters have to be found in a multidimensional deposition space, optimizing TCO for solar cells constitutes a very complex problem and deserves further research, including ideally a simultaneous up-scaling of the research results to industrially relevant substrate sizes. The important TCO semiconductors are impurity-doped ZnO, In_2O_3, SnO_2, and CdO, as well as the ternary compounds Zn_2SnO_4, $ZnSnO_3$, $Zn_2In_2O_5$, $Zn_3In_2O_6$, In_2SnO_4, $CdSnO_3$, and multi-component oxides consisting of combinations of ZnO, In_2O_3, and SnO_2. Sn doped In_2O_3 (ITO) and F doped SnO_2 TCO thin films are the preferable materials for most present applications. This study is focused on indium tin oxide.

2.5.3   ITO (Indium Tin Oxide)

Indium tin oxide is an n-type semiconductor with the band gap of more than 4 eV [49] and it is one of the most intensively studied TCO materials which has good characteristics such as high optical transparency, good electrical properties (such as carrier concentration, electrical mobility, and resistivity $\sim 10^{-4}$ Ω/cm), excellent substrate adherence, high thermal stability, and chemical inertness. Because of these optical and electrical performances, ITO films had been applied in different fields including solar cells, inorganic and organic light emitting devices, liquid crystal displays, laser diodes, and other optoelectronic devices. Hitherto, many techniques had been developed to prepare ITO thin films [50, 51].

Depending on the parameters which are applied during fabrication, different properties of ITO will be demonstrated. ITO depositions are highly sensitive in the oxygen flow range, sputtering pressure, sputtering time, sputtering power, sputtering temperature, and target which are called experimental details [51]. So, finding the right parameters is very critical.

Figure 2.13 shows the transmittance spectra in the wavelength between 300 and 1000 nm of 160 nm thick ITO films. The transmittance was calculated by subtracting absorption and reflection of the glass slides. Transmittance increased as substrate temperature increased. As the spectrum of ITO film prepared with a substrate temperature of 100 °C, the average transmittance of the wavelength of 400–1000 nm was low of 65.1%. For the sample prepared with a substrate temperature of 170 °C, average transmittance was 88.3%. Further increasing the substrate temperature to 300 °C increased the average transmittance to 94.5% [52].

The sheet resistance and band gap of the ITO films are functions of substrate temperature (Fig. 2.14). The sheet resistance of ITO film was 43 Ω/cm at a substrate temperature of 100 °C, decreased as substrate temperature increased, reached a

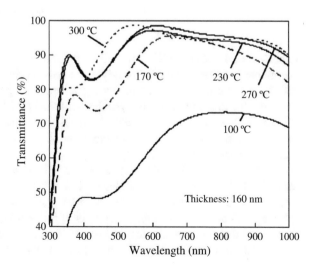

Fig. 2.13 Transmittance spectra of ITO films prepared at different substrate temperatures

Fig. 2.14 Sheet resistance and energy band gap properties of ITO thin films prepared at different substrate temperatures

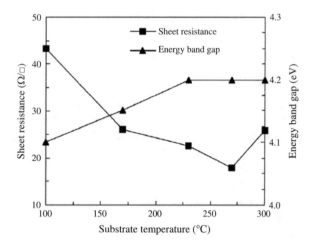

Fig. 2.15 Reflectivity of substrates and solar cell

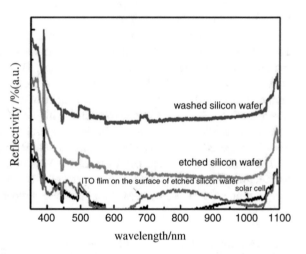

minimum of 17 Ω/cm at 270 °C, and then increased to 25 Ω/cm at 300 °C. The energy band gap of ITO films strongly depended on substrate temperatures of 100–230 °C, and was increased with increasing substrate temperature (Fig. 2.14). This is basically due to the increase in crystallinity of the ITO films. The energy band gap of the ITO films was almost constant when substrate temperature was 230–300 °C. A lower deposition of substrate temperature, <230 °C, increased the resistance and reduced the band gap of films [52].

Figure 2.15 shows the reflectivity of washed silicon substrate, etched silicon substrate, ITO film on the surface of etched silicon substrate, and nanocrystalline silicon solar cell, respectively. As can be seen from Fig. 2.15, the etched silicon substrate was able to reduce the reflectivity of incident light more than the washed silicon substrate; meanwhile, the ITO film on the surface of the etched silicon substrate reduced the reflectivity of incident light much more clearly. Therefore, the

ITO films on the surface of etched silicon substrate can be used as the window layer to reduce the incident light reflection [51]. The roughness of ITO is an issue for saving light and it is increased as substrate temperature increased [52].

References

1. P. Jackson, D. Hariskos, E. Lotter, S. Paetel, R. Wuerz, R. Menner, W. Wischmann, M. Powalla, New world record efficiency for Cu (In, Ga) Se2 thin-film solar cells beyond 20%. Prog. Photovolt. Res. Appl. **19**, 894–897 (2011)
2. A. Haugeneder, M. Neges, C. Kallinger, W. Spirkl, U. Lemmer, J. Feldmann, U. Scherf, E. Harth, A. Gügel, K. Müllen, Exciton diffusion and dissociation in conjugated polymer/ fullerene blends and heterostructures. Phys. Rev. B **59**, 15346 (1999)
3. S. Chih-Tang, R.N. Noyce, W. Shockley, Carrier generation and recombination in P-N junctions and P-N junction characteristics. Proc. IRE **45**, 1228–1243 (1957)
4. T. Tiedje, E. Yablonovitch, G.D. Cody, B.G. Brooks, Limiting efficiency of silicon solar cells. IEEE Trans. Electron Devices **31**, 711–716 (1984)
5. H. Schlangenotto, H. Maeder, W. Gerlach, Temperature dependence of the radiative recombination coefficient in silicon. Phys. Status Solidi A **21**, 357–367 (2006)
6. M. Zeman, *Introduction to Photovoltaic Solar Energy Solar Cells*, Collegemateriaal, 2003
7. M. Law, L.E. Greene, J.C. Johnson, R. Saykally, P. Yang, Nanowire dye-sensitized solar cells. Nat. Mater. **4**, 455–459 (2005)
8. M.R. Patel, *Wind and Solar Power Systems: Design, Analysis, and Operation* (CRC, Boca Raton, 2005)
9. G. Knier, How do photovoltaics work? Science@ NASA, 2002
10. J. Peet, J. Kim, N.E. Coates, W.L. Ma, D. Moses, A.J. Heeger, G.C. Bazan, Efficiency enhancement in low-bandgap polymer solar cells by processing with alkane dithiols. Nat. Mater. **6**, 497–500 (2007)
11. L. Zang, *Energy Efficiency and Renewable Energy Through nanotechnology* (Springer Verlag, London, 2011)
12. M.A. Green, K. Emery, Y. Hishikawa, W. Warta, Solar cell efficiency tables (version 37). Prog. Photovolt. Res. Appl. **19**, 84–92 (2011)
13. L. Kazmerski, NREL compilation of best research solar cell efficiencies. National Renewable Energy Laboratory, National Renewable Energy Laboratory (NREL), 2010
14. M.M.Y. Chan, C.H. Tao, V.W.W. Yam, Overview and highlights of WOLEDs and organic solar cells: from research to applications, in *WOLEDs and Organic Photovoltaics*, (Springer, Berlin, 2010), pp. 1–35
15. J. Zhao, A. Wang, M.A. Green, F. Ferrazza, 19.8% efficient "honeycomb" textured multicrystalline and 24.4% monocrystalline silicon solar cells. Appl. Phys. Lett. **73**, 1991 (1998)
16. O. Schultz, S. Glunz, G. Willeke, SHORT COMMUNICATION: ACCELERATED PUBLICATION: Multicrystalline silicon solar cells exceeding 20% efficiency. Prog. Photovolt. Res. Appl. **12**, 553–558 (2004)
17. R. Bergmann, C. Berge, T. Rinke, J. Schmidt, J. Werner, Advances in monocrystalline Si thin film solar cells by layer transfer. Sol. Energy Mater. Sol. Cells **74**, 213–218 (2002)
18. M.J. Keevers, T.L. Young, U. Schubert, M.A. Green, in *10% efficient CSG minimodules*. Proceedings of the 22 European Photovoltaic Solar Energy Conference, Milan, Italy, 2007, pp. 1783–1790
19. greentechsolar, http://www.greentechmedia.com/articles/read/stealthy-alta-devices-next-gen-pv-challenging-the-status-quo/
20. R. Venkatasubramanian, B. O'Quinn, J. Hills, P. Sharps, M. Timmons, J. Hutchby, H. Field, R. Ahrenkiel, B. Keyes, in *18.2% (AM1. 5) efficient GaAs solar cell on optical-grade*

polycrystalline Ge substrate. Conference Record of the Twenty Fifth IEEE Photovoltaic Specialists Conference (IEEE, 1996), pp. 31–36

21. C. Keavney, V. Haven, S. Vernon, in *Emitter structures in MOCVD InP solar cells*. IEEE Conference on Photovoltaic Specialists, vol. 141 (IEEE, 1990), pp. 141–144

22. I. Repins, M. Contreras, M. Romero, Y. Yan, W. Metzger, J. Li, S. Johnston, B. Egaas, C. DeHart, J. Scharf, in *Characterization of 19.9%-efficient CIGS absorbers*, 33rd IEEE Photovoltaic Specialists Conference (IEEE, 2008), pp. 1–6

23. I. Repins, M.A. Contreras, B. Egaas, C. DeHart, J. Scharf, C.L. Perkins, B. To, R. Noufi, 19 9%-efficient ZnO/CdS/CuInGaSe2 solar cell with 81·2% fill factor. Prog. Photovolt. Res. Appl. **16**, 235–239 (2008)

24. J. Kessler, M. Bodegard, J. Hedstrom, L. Stolt, New world record Cu (In, Ga) Se2 based mini-module: 16.6%, 2000, pp. 2057–2060

25. X. Wu, J. Keane, R. Dhere, C. DeHart, A. Duda, T. Gessert, S. Asher, D. Levi, P. Sheldon, *16.5%-Efficient CdS/CdTe Polycrystalline Thin-Film Solar Cell* (James & James Ltd., London, 2001)

26. S. Benagli, D. Borrello, E. Vallat-Sauvain, J. Meier, U. Kroll, J. Hötzel, J. Bailat, J. Steinhauser, M. Marmelo, G. Monteduro, in *High-efficiency amorphous silicon devices on LPCVD-ZnO TCO prepared in industrial KAI TM-M R&D reactor*. Proceedings of the 24th European Photovoltaic Solar Energy Conference (2009), pp. 2293–2298

27. K. Yamamoto, M. Yoshimi, Y. Tawada, Y. Okamoto, A. Nakajima, S. Igari, Thin-film poly-Si solar cells on glass substrate fabricated at low temperature. Appl. Phys. A Mater. Sci. Process. **69**, 179–185 (1999)

28. Y. Chiba, A. Islam, K. Kakutani, R. Komiya, N. Koide, L. Han, in *High efficiency dye sensitized solar cells*, Technical Digest. 15th International Photovoltaic Science and Engineering Conference, Shanghai (2005), pp. 665–666

29. M. Morooka, R. Ogura, M. Orihashi, M. Takenaka, Development of dye-sensitized solar cells for practical applications. Electrochemistry **77**, 960–965 (2009)

30. Konarka, http://www.konarka.com/

31. Solarmer, http://www.solarmer.com/

32. K. Mitchell, C. Eberspacher, J. Ermer, D. Pier, in *Single and tandem junction CuInSe2 cell and module technology*. Conference Record of the Twentieth IEEE Photovoltaic Specialists Conference, vol. 1382 (1988), pp. 1384–1389

33. J. Bailat, L. Fesquet, J. Orhan, Y. Djeridane, B. Wolf, P. Madliger, J. Steinhauser, S. Benagli, D. Borrello, L. Castens, Recent developments of high-efficiency micro-morph® tandem solar cells in Kai-M PECVD reactors, 2010, pp. 2340–2343. https://doi.org/10.4229/26thEUPVSEC2011-3BO.2.6

34. M. Yoshimi, T. Sasaki, T. Sawada, T. Suezaki, T. Meguro, T. Matsuda, K. Santo, K. Wadano, M. Ichikawa, A. Nakajima, in *High efficiency thin film silicon hybrid solar cell module on 1 m2-class large area substrate*. Proceedings of 3rd World Conference on Photovoltaic Energy Conversion, vol. 1562 (IEEE, 2003), pp. 1566–1569

35. Heliatek, http://www.heliatek.com/

36. D. Spitzer, Lattice thermal conductivity of semiconductors: a chemical bond approach. J. Phys. Chem. Solids **31**, 19–40 (1970)

37. Y. Xiong, Y. Xie, G. Du, H. Su, From 2D framework to quasi-1D nanomaterial: preparation, characterization, and formation mechanism of Cu3SnS4 nanorods. Inorg. Chem. **41**, 2953–2959 (2002)

38. P. Fernandes, P. Salomé, A. Da Cunha, A study of ternary Cu2SnS3 and Cu3SnS4 thin films prepared by sulfurizing stacked metal precursors. J. Phys. D. Appl. Phys. **43**, 215403 (2010)

39. M. Bouaziz, M. Amlouk, S. Belgacem, Structural and optical properties of Cu2SnS3 sprayed thin films. Thin Solid Films **517**, 2527–2530 (2009)

40. M. Bouaziz, J. Ouerfelli, S. Srivastava, J. Bernède, M. Amlouk, Growth of Cu2SnS3 thin films by solid reaction under sulphur atmosphere. Vacuum **85**, 783–786 (2011)

41. C. Wu, Z. Hu, C. Wang, H. Sheng, J. Yang, Y. Xie, Hexagonal CuSnS with metallic character: another category of conducting sulfides. Appl. Phys. Lett. **91**, 143104 (2007)
42. X. Chen, X. Wang, C. An, J. Liu, Y. Qian, Preparation and characterization of ternary Cu–Sn–E (E= S, Se) semiconductor nanocrystallites via a solvothermal element reaction route. J. Cryst. Growth **256**, 368–376 (2003)
43. X. Chen, H. Wada, A. Sato, M. Mieno, Synthesis, electrical conductivity, and crystal structure of Cu4Sn7S16 and structure refinement of Cu2SnS3. J. Solid State Chem. **139**, 144–151 (1998)
44. M. Bouaziz, J. Ouerfelli, M. Amlouk, S. Belgacem, Structural and optical properties of Cu3SnS4 sprayed thin films. Phys. Status Solidi A **204**, 3354–3360 (2007)
45. L.M. de Chalbaud, G.D. De Delgado, J. Delgado, A. Mora, V. Sagredo, Synthesis and single-crystal structural study of Cu2GeS3. Mater. Res. Bull. **32**, 1371–1376 (1997)
46. M. Onoda, X. Chen, A. Sato, H. Wada, Crystal structure and twinning of monoclinic Cu2SnS3. Mater. Res. Bull. **35**, 1563–1570 (2000)
47. M. Guittard, S. Benazeth, S. Jaulmes, M. Palazzi, P. Laruelle, J. Flahaut, Oxysulfides and oxyselenides in sheets, formed by a rare earth element and a second metal. J. Solid State Chem. **51**, 227–238 (1984)
48. J. Müller, B. Rech, J. Springer, M. Vanecek, TCO and light trapping in silicon thin film solar cells. Sol. Energy **77**, 917–930 (2004)
49. F.J. Haug, R. Biron, G. Kratzer, F. Leresche, J. Besuchet, C. Ballif, M. Dissel, S. Kretschmer, W. Soppe, P. Lippens, Improvement of the open circuit voltage by modifying the transparent indium–tin oxide front electrode in amorphous n–i–p solar cells. Prog. Photovolt. Res. Appl. **20**, 727 (2012)
50. A. Chebotareva, G. Untila, T. Kost, S. Jorgensen, A. Ulyashin, ITO deposited by pyrosol for photovoltaic applications. Thin Solid Films **515**, 8505–8510 (2007)
51. Z. Erjing, Z. Weijia, L. Jun, Y. Dongjie, H.J. Jacques, Z. Jing, Preparation of ITO thin films applied in nanocrystalline silicon solar cells. Vacuum **86**, 290 (2011)
52. S.Y. Lien, Characterization and optimization of ITO thin films for application in heterojunction silicon solar cells. Thin Solid Films **518**, S10–S13 (2010)

Chapter 3
Copper Tin Sulfide (CU_2SnS_3) Solar Cell Structures and Implemented Methodology

3.1 Introduction

Because very complex processes are involved in solar cells nowadays, scientists are thinking about software to predict certain factors in order to ensure that the best results can be achieved for their production. Several types of software are available, based on numerical simulation, and they are thus essential parts of the design of solar cell thin films. The Solar Cell Capacitance Simulator (SCAPS) is used for simulation of solar cells. The objective of SCAPS is to show how material properties (the bandgap, affinity, doping, mobilities, gap state defect distributions in the bulk and at interfaces, etc.) and the device design/structure together control the device physics and thereby also control the device responses to light, impressed voltage, and temperature. SCAPS allows users to learn the "whys" of device response to a given situation (that is, light bias, voltage bias, and temperature) through exploration and comparison of band diagrams, current components, recombinations, generation, and electric field plots available from AMPS (Analysis of Microelectronic and Photonic Structures) as a function of light intensity, voltage, and temperature.

From the solution provided by a SCAPS simulation, output such as the current voltage characteristics in the dark—and, if desired, under illumination—can be obtained. These may be computed as a function of temperature. For solar cell and detector structures, collection efficiencies as a function of voltage, light bias, and temperature can also be obtained. In addition, important information such as electric field distributions, free and trapped carrier populations, recombination profiles, and individual carrier current densities as a function of position can be extracted from the SCAPS program. As stated earlier, the versatility of SCAPS means it can be used to analyze transport in a wide variety of device structures, which can contain combinations of crystalline, polycrystalline, or amorphous layers. SCAPS is

© The Author(s), under exclusive license to Springer Nature Switzerland AG 2019
I. S. Amiri, M. Ariannejad, *Introducing CTS (Copper-Tin-Sulphide) as a Solar Cell by Using Solar Cell Capacitance Simulator (SCAPS)*, SpringerBriefs in Electrical and Computer Engineering, https://doi.org/10.1007/978-3-030-17395-1_3

formulated to analyze, design, and optimize structures intended for microelectronic, photovoltaic, or opto-electronic applications.

SCAPS is a Windows application program [1, 2], developed at the University of Ghent with LabWindows/CVI from National Instruments. It has been made available to university researchers in the photovoltaic community since the second Photovoltaic World Conference in Vienna in 1998. A solar cell simulation problem is stored in an ASCII file, which can be read and completely edited using the graphical user interface (GUI) of SCAPS.

The program is organized into a number of panels (or windows or pages, in other jargon) in which the user can set parameters or in which results are shown. The program opens with an action panel where the user can set an operating point (temperature, voltage, frequency, illumination) and an action list of calculations to carry out (I–V, C–V, C–f, $Q(\lambda)$). In each calculation the running parameter (V, f, or λ) is varied in the specified range, while all other parameters have the value specified at the operation point.

One can also load or edit a problem file. The user can navigate to many auxiliary panels. One, with the parameters for one defect level, is in one layer set. Also, the user can directly view previously calculated results (I–V, C–V, C–f, $Q(\lambda)$) and also band diagrams, electric fields, carrier densities, and partial recombination currents pertaining to a single mechanism. All calculations can be saved in an ASCII file, to be handled by the user with their preferred toolkit. Electrostatic potential and the two Fermi levels can now be edited by the user.

- When divergence occurs, the points calculated so far are not lost, but are shown.
- A status line informs the user about the progress of the calculation and the status after its termination.
- The possibility of illumination from either the p side or the n side is now built in.
- The open-circuit voltage (V_{oc}), short-circuit current density (J_{sc}), fill factor (FF), and device efficiency (η) parameters are calculated by interpolation of the calculated $I \pm V$ points.
- Many graphs can be shown on a logarithmic scale or a linear scale, by a click.
- After calculation, the data can be displayed on-screen, by a simple click. Parts of the data can be cut and pasted into other application programs, such as spreadsheets.
- The quality of the graph printing has been greatly improved.

3.2   Cell Model

In principle, any numerical program capable of solving the basic semiconductor equations could be used for modeling thin-film solar cells. The basic equations are Poisson's equation (relating the charge to the electrostatic potential Φ) and the continuity equations for electrons and holes. In one dimension, the total cell length L is divided into N intervals, and the values of Φ_i and of the electron and hole concentrations n_i and p_i at each of the intervals constitute the unknowns of the problem. They can be found by numerically solving 3N nonlinear equations—that is, the basic equations at each of the intervals i. Alternatively, one can choose F_i, E_{Fni}, and E_{Fpi} as independent variables instead of (Φ_i, n_i, p_i). Here, E_{Fn} and E_{Fp} are the quasi-Fermi energy levels for electrons and holes, respectively. The basic equations are nonlinear because the continuity equations contain a recombination term that is nonlinear in n and p.

Several additional requirements must be met if the program is to simulate thin-film polycrystalline solar cells realistically. It must allow for multiple semiconductor layers; six layers are needed. At the interfaces between the layers, discontinuities in the energy bands E_c and E_v, and hence in the bandgap E_g, can be present, and interface recombination can occur. A solar cell simulation program should be able to handle this. It should also correctly treat the problem of recombination and charge in deep states in the bulk of the layers. It should be able to calculate and simulate the relevant electro-optical measurements commonly carried out on thin-film solar cells—that is, not only the I–V characteristics but also the spectral response $Q(\lambda)$ and the capacitance measurements C–V and C–f; preferably all of these should be included as functions of ambient temperature. Finally, it should provide convergence for at least the most common thin-film cell structures—say, at least for basic transparent conductive oxide/cadmium sulfide/cadmium telluride (TCO/CdS/CdTe) structures and copper indium gallium selenide/CdS/TCO (CIGS/CdS/TCO) structures in normal working conditions. This is sometimes a weak point of a program dedicated to simulation of silicon cells, because of the occurrence of large-bandgap materials in typical thin-film cells. We now discuss how these features are implemented in our simulation program, SCAPS [1].

3.2.1 Deep Bulk Levels

In each layer, the type (donor or acceptor) and density of one shallow level can be defined; it is completely ionized and does not contribute to the recombination. Also, up to three deep levels can be defined. Recombination in these levels and their occupation is described by the Shockley–Read–Hall (SRH) formalism, and charge is defined by the occupation of the level and its type (donor, acceptor, or "neutral"— that is, a hypothetical center carrying no charge). The levels can be energetically distributed in the forbidden zone (single level, uniform band, Gauß, or exponential tail). The concentration of the shallow or deep states can vary spatially (uniform, step, linear, or exponential). Auger recombination is not treated but is a straightforward extension.

Amphoteric states—that is, states with multiple charge and energy states—cannot be handled for the moment. Their implementation, however, is straightforward and would extend the applicability of our program to amorphous cells. Conduction through deep states (hopping) is not implemented.

3.2.2 Optical Generation

SCAPS can handle steady-state illumination by monochromatic light or by the light of a standard spectrum (AM1.5G and AM1.5D are implemented by default). The wavelength region can be limited to simulate the use of long-pass and short-pass filters. Above that, a "small signal" illumination can be added to simulate a spectral response measurement. An exponential absorption law, characterized by a few user parameters, is assumed for all semiconductor layers. Extension to a completely user-defined $\alpha(\lambda)$ law is straightforward. Implementation of optical layers, which do not have an electronic effect but introduce a user-defined transmission $T(\lambda)$, will occur in the near future. No interference phenomena or optical confinement (texture and ray tracing) is handled at present. The steady-state semiconductor equations are discretized according to the exponential fitted finite difference scheme and solved by the Gummel iteration scheme [3].

3.2.3 Physics of Device Transport, Concept of Continuity, and Poisson's Equation

There are many physical processes that happen in a solar cell during its operation. Physical processes—such as photogeneration of charge carriers due to optical absorption of photons, charge carrier movements due to diffusion, and drift due to the existence of a concentration gradient and built-in potential and recombination of charge carriers via various recombination mechanisms—happen simultaneously at different rates. Hence, to quantitatively determine the overall performance of solar cells, the net rate of each process must be calculated. To do this, a notion of continuity of charge carriers is introduced, which links all of the charge carrier movements to the resulting net current in the device in the form of the continuity equation for electrons in the conduction band and holes in the valence band. In the same sense, Poisson's equation links free carrier populations, trapped charge populations, and ionized dopant populations to the electrostatic field present in a material system. Hence, the continuity equation for holes and electrons, as well as Poisson's equation, are solved with appropriate boundary conditions by use of a numerical solution technique. The one-dimensional continuity equation for both types of charge carriers and Poisson's equation are given in Eqs. (3.1) and (3.2). During operation of a solar cell, a steady-state condition of the charge carriers is assumed, in which the charge carriers are assumed to be independent of time evolution (the rate of change of the free carrier concentration is zero with respect to time).

$$\frac{1}{q}\left(\frac{dJ_n}{dx}\right) = -G_{op}(x) + R(x) \tag{3.1}$$

$$\frac{d^2\psi}{dx^2} = -\frac{d\varepsilon}{dx} = -\frac{q}{\varepsilon_s(x)}\left(N_D(x) - N_A(x) + p(x) - n(x) + p_t(x) - n_t(x)\right) \quad (3.2)$$

$G_{op}(x)$ and $R(x)$ are the optical generation rate and recombination rate with one-dimensional spatial dependence. $R(x)$ is the total recombination rate resulting from both direct recombination (band-to-band) and SRH recombination. In Eq. (3.2), N_D and N_A are the ionized donor- and acceptor-like concentrations, p and n are the free hole and electron concentrations, and p_t and n_t are the trapped hole and electron concentrations. All of these parameters are also functions of position x.

3.3 Cell Structure

Narrow-bandgap copper tin sulfide (CTS) powder was synthesized by a simple ball milling process from the powdered elements copper (Cu, 99.9%; Wako Chemicals), tin (Sn, 99.5%; Aldrich), and sulfur (S, 99.9%; Kishida Chemicals) at a molar ratio of 2:1:3. The obtained CTS powder was dispersed by propylene glycol (Kando Chemicals) to prepare a paste. The paste was then deposited on preprepared fluoride-doped tin oxide (FTO) glass/TiO$_2$/In$_2$S$_3$ substrates by a doctor blade method (both the compact TiO$_2$ window layer and the In$_2$S$_3$ buffer layer were prepared by a conventional spray pyrolysis method). After being dried in air at 125 °C for 5 min, the samples (FTO glass/TiO$_2$/In$_2$S$_3$/CTS) were annealed in N$_2$ ambience to improve the crystallinity of the CTS. Finally, Mo electrodes were sputtered onto the annealed samples to complete the CTS solar cell fabrication (FTO glass/TiO$_2$/In$_2$S$_3$/CTS/Mo). The active area of the solar cell was 0.5 cm × 0.5 cm. To confirm the crystallinity and phase composition of the CTS absorber layer, x-ray diffraction (XRD) measurements were performed [4]. Figure 3.1 depicts the XRD patterns of the CTS absorber layers prepared at various annealing temperatures.

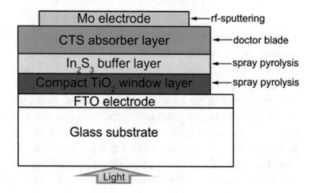

Fig. 3.1 Structure of a solar cell. *CTS* copper tin sulfide, *FTO* fluoride-doped tin oxide, *rf* radiofrequency

Table 3.1 Overall semiconductor material properties

Abbreviation	Material parameter, symbol (unit)
EPS	Relative permittivity, ε_r
MDN	Electron band mobility, μ_n (cm^2/V.s)
MDP	Hole band mobility, μ_p (cm^2/V.s)
NA	Acceptor concentration, N_a (1/cm^3)
ND	Donor concentration, N_d (1/cm^3)
EG	Electric bandgap, E_g (eV)
NC	Effective density of states in the conduction band, N_c (1/cm^3)
NV	Effective density of states in the valence band, N_v (1/cm^3)
CHI	Electron affinity, χ (eV)
Φ_b (eV)	Barrier height
S	Surface recombination velocity

3.4   SCAPS Numerical Modeling Computer Program

SCAPS is a computer program capable of modeling solar cell physical and electronic structures and numerically solving them to give performance parameter output in terms of the V_{oc}, J_{sc}, FF, and η values. This simulator has all of the algorithms needed to calculate the basic semiconductor properties—as well as solving the continuity and Poisson's equations—with an interactive and user-friendly GUI. Various different structures of solar cells can be modeled in SCAPS. The key to "good" numerical modeling of a solar cell lies in the input of the material property parameters by the user. The closer the material property parameter input is to the actual value in the real world, the more relevant the results will be to the practical implementation of the solar cell. The overall material property parameters that are needed are listed in Table 3.1, with the numerical values that are used.

3.5   Cu$_2$SnS$_3$ Solar Cell Structure Parameters

With this software we can retrieve all of the data about all of the layers for CTS by clicking on the button showing the name of each layer: "p-CTS" (p-type copper tin sulfide), "n-ZnO" (n-type zinc oxide), or "n-CdS" (n-type cadmium sulfide). With one click, all of the parameters will be displayed, as shown in Fig. 3.2, which lists the parameters of the layers with their values. Some values can be changed just by clicking. There is one panel listing the names of defects, and in this panel the defect types that the layer possesses can be defined. The absorption layer is untouched; some parameters are changed according to the properties of the material. The data must be changed to yield the best result in this simulation. After all of the parameters are changed, the "Save" button must be clicked to save the structure so we can also use these data later on.

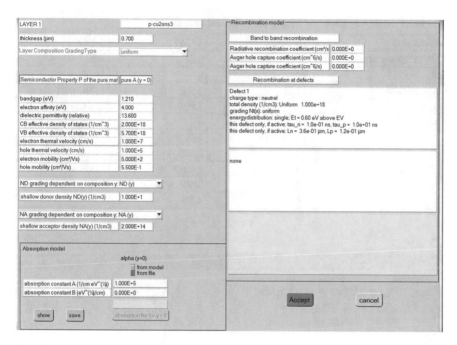

Fig. 3.2 Copper tin sulfide (*CTS*) cell parameters panel

3.6   Setting Up a Reference Case and Parameters in the Numerical Simulation

On the basis of ZnO/CdS/CTS, a solar cell structure is selected and the relevant data are taken from experimental data based on known solar cell theory. The V_{oc}, J_{sc}, FF, and η values obtained from the simulation are then used as a basis for comparing similar simulated values from other variations of the CTS-based solar cell [5, 6]. The base parameters that are used to model the individual layers that constitute the reference solar cell in this simulation are listed in Table 3.2 [5].

ε permittivity, μ_n electron band mobility, μ_p hole band mobility, χ electron affinity, *CTS* copper tin sulfide, E_g bandgap, *NA* acceptor concentration, N_c effective density of states in the conduction band, *n-CdS* n-type cadmium sulfide, *ND* donor concentration, N_v effective density of states in the valence band, *n-ZnO* n-type zinc oxide, *p-CTS* p-type copper tin sulfide

3.7   Batch Operation (Calculation)

In the SCAPS software a batch operation is a specific function that can be referred to as a batch calculation. By use of this function, a simulation can be run in which the values of one or more parameters are varied numerically to obtain the

Table 3.2 Material parameters used in the numerical analysis

Parameter	n-ZnO	n-CdS	p-CTS
Layer thinness (μm)	0.20	0.05	2.20
$\varepsilon/\varepsilon_0$	9	9	10.00
μ_n (cm²/V.s)	100	3.50E + 2	1.00E + 2
μ_p (cm²/V.s)	25	5.00E + 1	2.50E + 1
NA (1/cm³)	0	0	2.00E + 14
ND (1/cm³)	1.0E + 18	1.0E + 17	1.00E + 1
E_g (eV)	3.3	2.410	1.210
N_c (1/cm³)	2.20E + 18	1.80 + 19	2.20E + 18
N_v (1/cm³)	1.80E + 19	2.40E + 18	1.80E + 19
χ (eV)	4.6	4.50	4

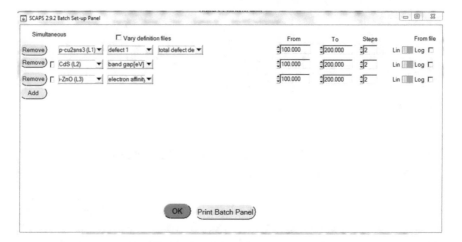

Fig. 3.3 Batch set-up panel

performance result for the solar cell. For setting up the batch calculation, the "Batch Set-up" button must be clicked in the main action panel of SCAPS (Fig. 3.3). A new batch input panel will then pop up, and the user can change the parameters, depending on the material properties, which can be varied. Some parameters can be changed just by clicking on the "Add" button in the panel and specifying several values. In the "Value" box, the particular parameters can be inserted that are to be used in the simulation.

After the values of the relevant parameters have been changed, the "OK" button can be clicked to return to the main panel of SCAPS. The batch calculation can be started before the simulation for the batch calculation of the solar cell, as shown in Fig. 3.4.

Fig. 3.4 Checking the "Do Batch Calculation" box

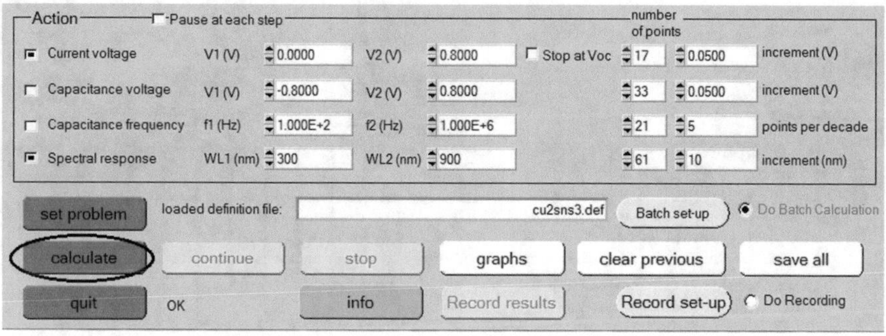

Action	Pause at each step						number of points		
☑ Current voltage	V1 (V)	0.0000	V2 (V)	0.8000	☐ Stop at Voc		17	0.0500	increment (V)
☐ Capacitance voltage	V1 (V)	-0.8000	V2 (V)	0.8000			33	0.0500	increment (V)
☐ Capacitance frequency	f1 (Hz)	1.000E+2	f2 (Hz)	1.000E+6			21	5	points per decade
☑ Spectral response	WL1 (nm)	300	WL2 (nm)	900			61	10	increment (nm)

set problem | loaded definition file: | cu2sns3.def | Batch set-up | ⦿ Do Batch Calculation

calculate | continue | stop | graphs | clear previous | save all

quit | OK | info | Record results | Record set-up | ○ Do Recording

Fig. 3.5 Simulation execution panel

3.8 Obtaining the Results

By clicking on the "Calculate" button after all relevant procedures have been done, the simulation can be run, as shown in Fig. 3.5.

When the "calculate" button is clicked, the energy band panel will pop up, and it has four significant graphs. These graphs are the carrier density, energy band diagram, occupation probability, and current density of deep defects. In Fig. 3.6, the energy band panel is shown.

Right after the simulation, the results for the I–V characteristics and the quantum efficiency (QE) of the solar cell can be considered by clicking on the "QE" and "I–V" buttons. The I–V panel will appear with the characteristic results (for example, the V_{oc}, J_{sc}, FF, and η values). In Fig. 3.7, the I–V panel is shown.

For QE, the same procedure is followed. To view the QE result for the solar cell, the "QE" button can be clicked to display a new panel, which shows the spectral response of the solar cell. By use of the "save" button in the I–V panel, the result will be saved and can be analyzed later on, as shown in Fig. 3.8. To understand more about these data, we can put all of the data into the Excel software program to demonstrate diagrams and curves. To do this, at each step after getting the I–V and QE information, by clicking on the "show" button, the data will appear that can be transferred to Excel. When all of these procedures have been performed and the data have been transferred to Excel, it is possible to draw all of the curves and diagrams to provide good visualization of the results and the cell performance. In this area, comparison of the results can be achieved. In the following Chap. 4, all of the results and the curves and diagrams are discussed.

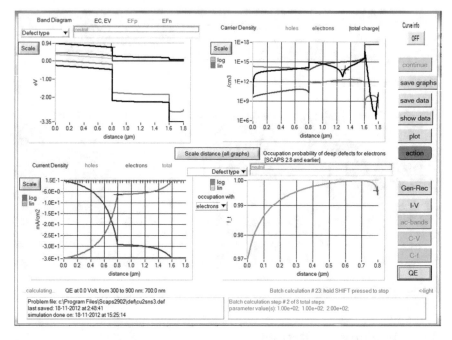

Fig. 3.6 Energy band panel

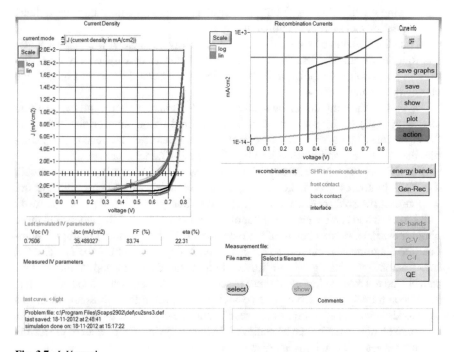

Fig. 3.7 *I–V* panel

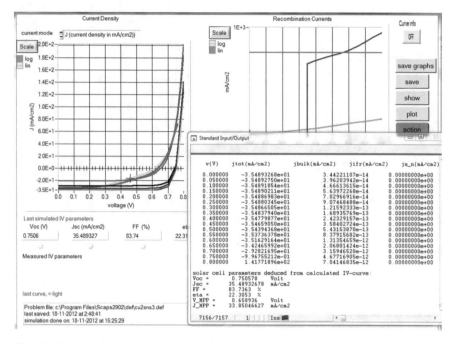

Fig. 3.8 Simulation result data

References

1. M. Burgelman, P. Nollet, S. Degrave, Modelling polycrystalline semiconductor solar cells. Thin Solid Films **361**, 527–532 (2000)
2. A. Niemegeers, S. Gillis, M. Burgelman, in Interpretation of capacitance spectra in the special case of novel thin film CdTe/CdS and CIGS/CdS solar cell device structures, 2nd World Conference on Solar Energy Conversion, 1998
3. S. Selberherr, *Analysis and simulation of semiconductor devices* (Springer-Verlag, New York, 1984)
4. Q. Chen, X. Dou, Y. Ni, S. Cheng, S. Zhuang, Study and enhance the photovoltaic properties of narrow-bandgap Cu_2SnS_3 solar cell by p–n junction interface modification. J. Colloid Interface Sci. **376**, 327–330 (2012)
5. N. Amin, M.I. Hossain, P. Chelvanathan, A.S.M.M. Uzzaman, K. Sopian, in Prospects of Cu_2ZnSnS_4(CZTS) solar cells from numerical analysis, 2010 International Conference on Electrical and Computer Engineering (ICECE), 2010, pp. 730–733
6. A. Yildirim, S.T. Mohyud-Din, D.H. Zhang, Analytical solutions to the pulsed Klein–Gordon equation using modified variational iteration method (MVIM) and Boubaker polynomials expansion scheme (BPES). Comput. Math. Appl. **59**, 2473–2477 (2010)

Chapter 4
CTS Solar Cell Performance Analysis and Efficiency Characterizations

4.1 Introduction

Structural parameter variation of CTS solar cell has been studied by solar cell capacitance simulator (SCAPS) in terms of buffer layer and absorber layer thickness and band gap, effect of i-ZnO thickness on total efficiency of solar cell to find out the optimum electrical performance.

4.2 Effects of Various Layer Thicknesses of Cds Buffer Layer

To rationalize the simulation, the novel p-CTS heterojunction solar cells with CdS buffer layer have been verified in terms of buffer layer thickness. At the beginning of the simulation, the buffer layer thickness has been varied from 10 nm to 90 nm to carry out the optimum electrical performance of these heterojunction solar cells. In addition, 4.0 μm of CTS layer thickness is used during the simulation. It has been found that the efficiency of the solar cell is decreasing with the increase in CdS buffer layer thickness.

The maximum efficiency which has been found was 20.36%, when the buffer layer thickness is 10 nm. The best possible buffer layer thickness of CdS is between 10 nm and 30 nm. The observed efficiencies are 20.36% and 20.01% of the thickness of 10 nm and 30 nm. It can be related to thinner layer thickness. From diagrams it can be concluded that as the buffer layer will increase to 90 nm, the efficiency of the solar cell drops. This is due to photon loss that occurs on the buffer layer. $J–V$ characteristic of a solar cell with variable buffer layer thickness is given in Fig. 4.1

© The Author(s), under exclusive license to Springer Nature Switzerland AG 2019
I. S. Amiri, M. Ariannejad, *Introducing CTS (Copper-Tin-Sulphide) as a Solar Cell by Using Solar Cell Capacitance Simulator (SCAPS)*, SpringerBriefs in Electrical and Computer Engineering, https://doi.org/10.1007/978-3-030-17395-1_4

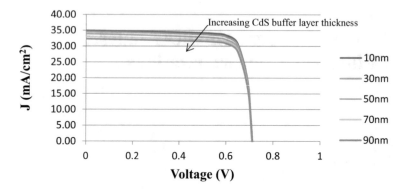

Fig. 4.1 *J–V* characteristic of solar cell with variable buffer layer thickness

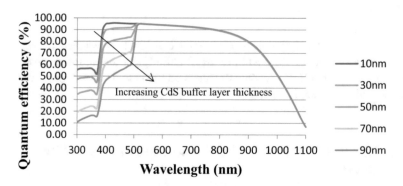

Fig. 4.2 Spectral response of solar cell with variable buffer layer thickness

and spectral response of cell is given in Fig. 4.2. Open circuit voltage (V_{oc}), short circuit current density (J_{sc}), fill factor, and efficiency of the CTS solar cells are shown in Fig. 4.3.

4.3 Effects of Various Layer Thicknesses of CTS Absorber Layer

To rationalize the simulation, the p-CTS absorber layer has been verified in terms of the layer thickness. At the beginning of the simulation, the absorber layer thickness has been varied from 1 μm to 4 μm, to carry out the optimum electrical performance of this heterojunction solar cell. It has been found that the efficiency of the solar cell is increased by the increase in thickness of absorber layer.

The highest efficiency 20.36% is achieved, when the absorber layer thickness is 4 μm. Efficiencies of 20.42% and 20.45% can be achieved by the thickness of 5 μm

Fig. 4.3 Cell
performances with the
variable thickness of CdS
buffer layer

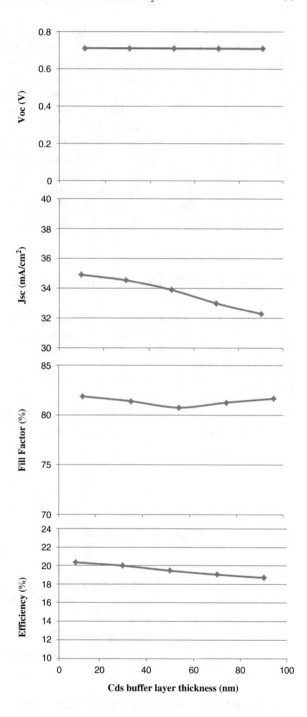

and 6 µm; however, the increasing of efficiency is low and insignificant. There are two factors that contribute to increase the efficiency when the absorber layer thickness is increased. Firstly, when the absorber layer increases, the minority carrier diffusion length also increases. This means electrons which are produced in p-type absorber layer due to the photon absorption have longer diffusion length. More photons with longer wavelength can be absorbed in the absorber layer and subsequently more electrons can be collected which will result in electricity generation. Secondly, when the absorber layer thickness is increased, the depletion region also increases. Hence, electrons can drift towards the n-type region due to the electric field produced by the depletion region easily. Open circuit voltage (V_{oc}), short circuit current density (J_{sc}), fill factor, and efficiency of the CTS solar cells are shown in Fig. 4.4. The J–V characteristics of the solar cells are shown in Fig. 4.5, and spectral response of the solar cell is shown in Fig. 4.6.

4.4   Effect of CTS Layer Band Gap on Cell Efficiency

In this work, the CTS layer band gap has been varied from 0.9 eV to 1.25 eV to see the effects on electrical performance. It has been found that while the band gap increases the efficiency of the solar cell decreases. In this simulation 0.9 eV band gap results in 11.58% cell efficiency, while by increasing the band gap value the efficiency of cell increases, as 1.25 eV band gap results in 21.96% cell efficiency. The J–V characteristics and spectral response of the cells are shown in Fig. 4.7. Figure 4.8 shows the cell performance with variable absorber layer band gap.

Band gap energy E_g: The absorption coefficient a is related [1, 2] to the incident photon energy $h\upsilon$ by the relation:

$$\frac{\sqrt[n]{\alpha h\upsilon}}{h\upsilon - E_g} = C \tag{4.1}$$

where E_g is the band gap, C is a constant, h is Planck constant, and n the index indicating the transition type:

$$n = \begin{cases} \frac{6}{3} \text{ for allowed} & \}\text{Indirect transition} \\ \frac{2}{3} \text{ for forbidden} & \\ \frac{1}{3} \text{ for allowed} & \}\text{direct transition} \\ \frac{2}{3} \text{ for forbidden} & \end{cases} \tag{4.2}$$

Measurements of the υ-dependent absorption coefficient α via transmittance–reflectance ($T(\lambda) - R(\lambda)$) allowed monitoring curves of $\sqrt[n]{\alpha h\upsilon}\,|_{n=2}$ versus $h\upsilon$ as guides to determine band gap energies E_g. The obtained values are gathered in Table 4.1 [3].

Fig. 4.4 Cell
performances with variable
thickness of CTS absorber
layer

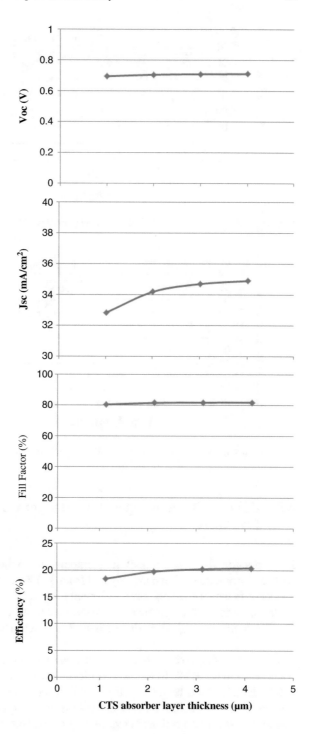

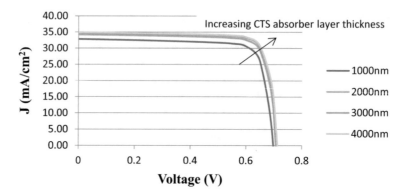

Fig. 4.5 *J–V* characteristics of cell with variable thickness of CTS absorber layer

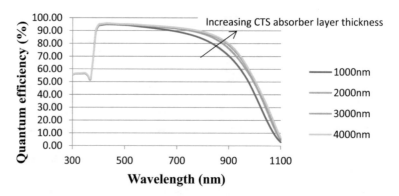

Fig. 4.6 Spectral response of the solar cell for different absorber layer thickness

4.5   Effect of Various Operating Temperature on Solar Cell Efficiency

In this section, the effect of operating temperature on the performance of the CTS solar cell structure has been investigated from 300 K to 400 K. When the operating temperature increases, electrons in solar cell gain an extensive energy, but instead of contributing to electricity generation, these electrons become unstable and recombine with the holes before the carriers could reach the depletion region and collected.

Efficiency of 20.36% is obtained for operating temperature of 300 K, as the temperature is increased to 310 K, the cell efficiency decrease to 19.69% and 19.02% for 320 K. Finally, it has been found that in 400 K the cell efficiency drops to 13.42%. Figures 4.9 and 4.10 show the *J–V* characteristics and spectral response of the cells. Figure 4.11 shows the cell performance with various operating temperature.

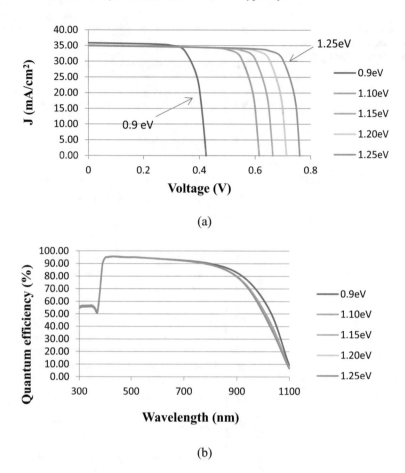

Fig. 4.7 (**a**) The *J–V* characteristics of cell, (**b**) spectral response of the cell with different absorber layer band gap

4.6   Effects of Various Layer Thicknesses of ZnO as N-Type Layer

To rationalize the simulation, the i-ZnO n-type layer has been verified in terms of the layer thickness for an n layer of p–n junction. At the beginning of the simulation, the n-layer thickness has been varied from 50 nm to 300 nm, to carry out the optimum electrical performance of this heterojunction solar cell. It has been found that the efficiency of the solar cell is increased by the increase in thickness of n junction. In the p–i–n junctions, the region of the internal electric field is extended by inserting an intrinsic, i, layer between the p-type and the n-type layers. The i-layer behaves like a capacitor and it stretches the electric field formed by the p–n junction across it. Electrons flow from the p-type into the n-type region and holes from the n-type

Fig. 4.8 Cell performances with variable absorber layer band gap

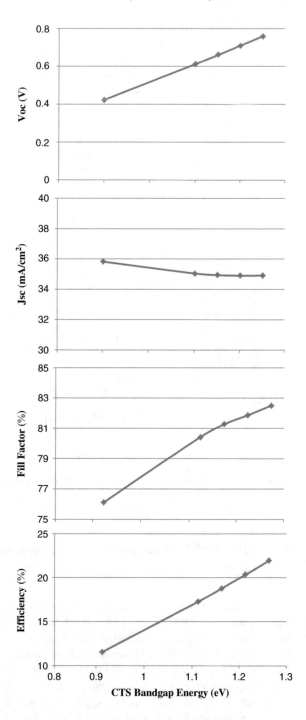

Table 4.1 χ-dependent band gap energy

χ	2.4	3.2	4.0	4.8
E_g(eV)	1.38	1.24	1.21	1.17

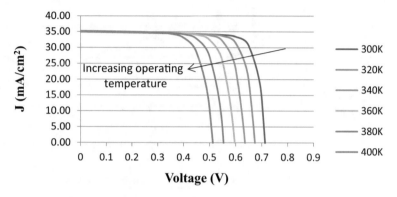

Fig. 4.9 *J–V* characteristics with various operating temperature (K)

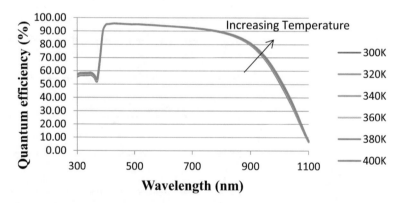

Fig. 4.10 Spectral responses with various operating temperature (K)

into the p-type region. The flow of the photo-generated carriers causes the so-called photo-generation current, J_{ph}. Forward bias, which adds to the thermal-generation current, J_{gen}. When no external contact between the n-type and the p-type regions is established, which means that the junction is in the open-circuit condition, no net current can flow inside the p–n junction.

The highest efficiency 20.36% is achieved, when the n-type layer thickness is 50 nm. Although efficiencies of 20.16% and 20.06% can be achieved by the thickness of 100 and 150 nm, the decrease in efficiency is small and insignificant. The region in the solar cell where the n-type and p-type layers meet is called the p–n junction. As may have already guessed, the p-type layer contains more positive

Fig. 4.11 CTS cell
performance with various
operating temperature (K)

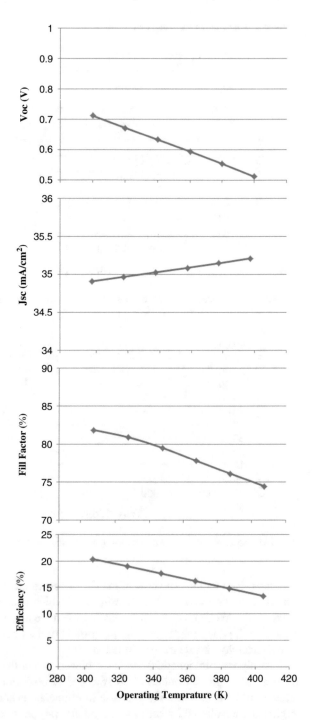

Fig. 4.12 Cell
performances with variable
thickness of ZnO n-layer

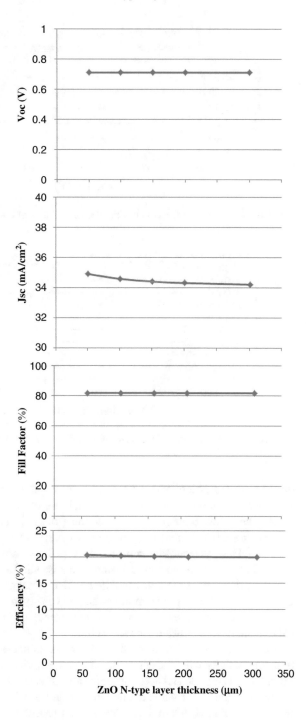

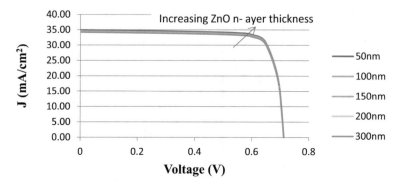

Fig. 4.13 *J–V* characteristics of cell with variable thickness of ZnO n-layer

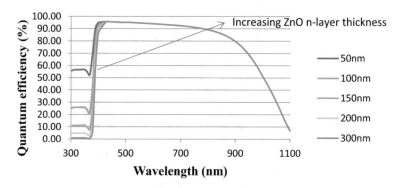

Fig. 4.14 Spectral response of the solar cell for different ZnO n-layer thickness

charges, called holes, and the n-type layer contains more negative charges, or electrons. When p-type and n-type materials are placed in contact with each other, current will flow readily in one direction (forward biased) but not in the other (reverse biased). An interesting interaction occurs at the p–n junction of a darkened solar cell. Extra valence electrons in the n-type layer move into the p-type layer filling the holes in the p-type layer forming what is called a depletion zone. The depletion zone does not contain any mobile positive or negative charges. Moreover, this zone keeps other charges from the p- and n-type layers from moving across it. So, to recap, a region depleted of carriers is left around the junction, and a small electrical imbalance exists inside the solar cell.

An example of such a material is a n-type semiconductor, in which a large electron conductivity with respect to the hole conductivity is caused namely by a large difference in electron and hole concentrations. Electrons can easily flow through the n-type semiconductor while the transport of holes, which are the minority carriers in such material, is due to the recombination processes very limited. The opposite hols for electrons in a p-type semiconductor, which is an example of the hole membrane. In order to minimize the injection of holes from the absorber into the n-type semiconductor an energy barrier should be introduced in the valence band, E_v, between the n-type semiconductor and the absorber. Ideally, this can be achieved by choosing an n-type semiconductor that has a larger band gap than that of the absorber and the energy difference is located in the valence band of the two materials. When the n-type layer increases, the minority carrier diffusion lengths also decrease. An electron can drift towards the n-type region due to the electric field produced by the depletion region easily. Open circuit voltage (V_{oc}), short circuit current density (J_{sc}), fill factor, and efficiency of the i-ZnO n-type layer are shown in Fig. 4.12. The $J-V$ characteristics of the solar cells are shown in Fig. 4.13, and spectral response of the solar cell is shown in Fig. 4.14.

References

1. A. Martínez, D. Acosta, A. López, Efecto del contenido de Sn sobre las propiedades físicas de películas delgadas de TiO2. Superficies vacío **16**, 5–9 (2003)
2. A. Sim, The dark-spot method for measuring the diffusion constant and length of excess charge carriers in semiconductors. Proc. IEE - Part B: Electron. Commun. Eng. **106**, 311–328 (1959)
3. D.H. Zhang, F.W. Li, Boubaker Polynomials Expansion Scheme (BPES) optimisation of copper tin sulfide ternary materials precursor's ratio-related properties. Mater. Lett. **64**, 778–780 (2010)

Chapter 5
A Summary of Semiconductor Solar Cells and Future Works

5.1 Summary

Thin-film solar cells (TFSCs) based on earth-abundant, low-cost, and nontoxic materials deposited by nonvacuum liquid processes have the potential to solve the present and future electric power crisis. The heart of a TFSC is the absorber layer, which efficiently harvests sunlight. This is ideally a p-type semiconductor with a direct bandgap (E_g) of ~1.3 eV and an optical absorption coefficient (α_{opt}) of 10^5 cm^{-1}. Copper tin sulfide (Cu$_2$SnS$_3$; CTS) is a p-type semiconductor with an E_g of ~1.1 eV and an $\alpha_{opt} > 10^5$ cm^{-1}; thus, it qualifies as a prospective absorber material. Structural variation of CTS solar cells has been studied in terms of the buffer layer and absorber layer thickness, the bandgap effect on the conversion efficiency of the cell, and the thickness of i-ZnO as an n-type layer. This study yielded the following conclusions:

First, it was found that the efficiency of the solar cell decreases with increased thickness of the CdS buffer layer. The highest efficiency (20.36%) is achieved when the buffer layer thickness is 10 nm. The best possible buffer layer thickness of CdS is between 10 nm and 30 nm. The observed efficiency values are 20.36% and 20.01% at thicknesses of 10 nm and 30 nm, respectively.

Secondly, in terms of the CTS absorber layer thickness, it was found that the efficiency of the solar cell increases with increased thickness of the absorber layer. At the very first simulation, the absorber layer thickness was varied from 1 μm to 4 μm to achieve the optimum electric performance of this heterojunction solar cell. The highest efficiency (20.36%) is achieved when the absorber layer thickness is 4 μm. However, efficiency values of 20.42% and 20.45% can be achieved with thicknesses of 1 μm and 3 μm, respectively, although the increase in efficiency is small and insignificant.

© The Author(s), under exclusive license to Springer Nature Switzerland AG 2019   63
I. S. Amiri, M. Ariannejad, *Introducing CTS (Copper-Tin-Sulphide) as a Solar Cell by Using Solar Cell Capacitance Simulator (SCAPS)*, SpringerBriefs in Electrical and Computer Engineering, https://doi.org/10.1007/978-3-030-17395-1_5

Thirdly, the CTS layer bandgap was varied from 0.9 eV to 1.25 eV to investigate the effects on electric performance. It was found that when the bandgap increases, the efficiency of the solar cell decreases. In this simulation a 0.9-eV bandgap resulted in 11.58% cell efficiency; with an increase in the bandgap value, the efficiency of the cell increases, and a 1.25-eV bandgap results in 21.96% cell efficiency.

Also, the effect of the operating temperature on the performance of the CTS solar cell structure was investigated from 300 K to 400 K. Efficiency of 20.36% is achieved at 300 K. As the temperature increases to 310 K and 320 K, the cell efficiency values decrease to 19.69% and 19.02%, respectively. Finally, it was found that at an operating temperature of 400 K, the cell efficiency drops to 13.42%.

Finally, this work included assessment of the i-ZnO n-type layer thickness. It was found that the efficiency of the solar cell is decreased by increased thickness of the n-layer. At the very first simulation, the absorber layer thickness was varied from 50 nm to 300 nm for optimum electric performance of this heterojunction solar cell. The highest efficiency (20.36%) is achieved when the absorber layer thickness is 50 nm. However, efficiency values of 20.16% and 20.06% can be achieved with thicknesses of 100 and 150 mm, respectively, although the increase in efficiency is small and insignificant.

5.2   Future Recommendations

Like Cu_2SnS_3 and SnS, the p-type semiconductor Cu_2SnS_3 consists of only earth-abundant and low-cost elements, and it shows comparable optoelectronic properties with respect to Cu_2SnS_3 and SnS, making it a promising candidate for photovoltaic applications.

The combination of CTS with CdS and ZnO yields the best results possible for this absorber material and allows a direct start without the need to spend too much time searching for a working device structure. Instead, one can concentrate on the properties of CTS and leave the subtleties of the solar cell structure for future work.

Index

© The Author(s), under exclusive license to Springer Nature Switzerland AG 2019
I. S. Amiri, M. Ariannejad, *Introducing CTS (Copper-Tin-Sulphide) as a Solar Cell
by Using Solar Cell Capacitance Simulator (SCAPS)*, SpringerBriefs in Electrical
and Computer Engineering, https://doi.org/10.1007/978-3-030-17395-1

Printed in the United States
By Bookmasters